T0231890

THE LAST ͅ

Opening the Door on the
Global Sanitation Crisis

Maggie Black and Ben Fawcett

publishing for a sustainable future

London • New York

First published by Earthscan in the UK and USA in 2008

ISBN: 978-1-84407-543-0 hardback
         978-1-84407-544-7 paperback

Typeset by MapSet Ltd, Gateshead

Cover design by Susanne Harris

For a full list of publications please contact:

Earthscan

2 Park Square, Milton Park, Abingdon, Oxon OX14 4RN

Simultaneously published in the USA and Canada by Earthscan

711 Third Avenue, New York, NY 10017

Earthscan in an imprint of the Taylor & Francis Group, an informa business

Earthscan publishes in association with the International Institute
for Environment and Development

A catalogue record for this book is available from the British Library

Library of Congress Cataloging-in-Publication Data

Black, Maggie, 1945-
  The last taboo : opening the door on the global sanitation crisis / Maggie Black and Ben
Fawcett.
       p. ; cm.
  Includes bibliographical references.
  ISBN 978-1-84407-543-0 (hardback) — ISBN 978-1-84407-544-7 (pbk.) 1. Sanitation. I.
Fawcett, Ben. II. Title.
  [DNLM: 1. Sewage. 2. Communicable Disease Control. 3. Feces—microbiology. 4. Toilet
Facilities. 5. Waste Management—history. 6. World Health. WA 778 B627L 2008]
  RA567.B63 2008
  363.72—dc22
                                                                              2007051581

'Great is sanitation; the greatest work, except discovery, I think, that one can do. What is the use of preaching high moralities, philosophies, policies and arts to people who dwell in appalling slums? You must wipe away those slums, that filth, these diseases. We must begin by being cleansers.'

Sir Ronald Ross, 1857–1932

Contents

List of Figures and Tables

Figures

Tables

Foreword

Walk through any of the developing world's shanty-towns or slums, and you are struck by the extraordinary efforts made by people living in the most squalid places on the planet to keep themselves clean and well groomed. Given their lack of bathing and laundry facilities, the mud and debris on footpaths, the rotting trash in open spaces, and the ramshackle state of much of their housing, you wonder not why some children are ill-kempt – because that is obvious – but how on earth so many of those living in such difficult circumstances manage to have clean shoes, fresh clothes and be so well turned-out. And you wonder how they manage to deal with their bodily needs.

This book makes us think about these things, and does so with great power, not in a way to feed our inhibitions, but to help us to overcome them. It also challenges our false assumptions about the lack of demand for sanitation services in the poorer parts of the developing world, and indicates what we can do to respond. In the International Year of Sanitation, this wake-up call is more than welcome – it is essential. Too often, the 'sanitation' component of 'water and sanitation services' is referred to only in passing, as if clean water alone will solve the personal environmental crisis confronting the world's poorest citizens. Here, the vital importance of sanitation in the public health revolution we so desperately need to enable us to meet not just one, but many, of the Millennium Development Goals is given what it deserves: pride of place.

At the outset of the 21st century, the lack of sanitation endured by at least 2.6 billion people – 40 per cent of the world's citizens – is a hidden international scandal. This, not lack of water, is the principal reason for the spread of diarrhoeal diseases and the toll they take on human lives – 2.2 million a year, mostly among children under the age of five. Yes, it is true that the infectious agents may be imbibed. But this is because so many faecal particles and pathogens are present in the environment, not contained hygienically in facilities for the purpose. As a result, they find their way onto hands, feet, faces, clothes and utensils, as well as into drinking water and food. Unfortunately, we are too squeamish to lay the blame for this disease load squarely where it belongs – on excreta. And on the need to manage it and dispose of it properly in the many urban and rural environments in which sewerage of the kind we all enjoy in the industrialized world is not feasible, for reasons of expense and impracticality.

Even those of us who recognize that major sanitary reform is required for better health in the developing world often fail to appreciate how undignified and personally distressing it is to have no decent place to 'go'. In the past, women in rural areas were able to go out in the cover of dark to a nearby area, set aside specially for the purpose, with bushes and trees to protect their privacy. Nowadays, the bushes and trees are gone, there are miles to walk, and girls and women risk being attacked if they venture out so far in the night-time. And for girls, especially once they have reached adolescence, the lack of facilities in schools is a frequent reason for parents to end their daughters' education: their propriety and modesty is disrespected and their sense of decency is denied.

At the very least, all governments should make it a target in this International Year of Sanitation to provide proper facilities in every school and child care centre, along with hygiene education in the classroom to make sure that toilet and washing blocks are used and well maintained. How would we feel if our children went off for the day to fulfil their right to education in a place where basic human needs and functions were ignored? In this context, I highly commend the work of UNICEF worldwide, an organization that has been instrumental in bringing this book to fruition; and of its partner in my own country, the IRC International Water and Sanitation Centre in Delft. School sanitation is one of the important approaches covered in these pages, and represents real hope for a healthier future for millions of the world's citizens.

The authors of this book deserve credit for bringing out into the open a subject we instinctively avoid. The story they tell of today's sanitary heroes and endeavours is truly compelling, just as compelling as the more famous story of 19th-century sanitary reform in the industrialized world. We need to recapture the energy and determination of that earlier generation so as to extend public health engineering of an appropriate and sustainable kind to the other half of humanity. It is my belief that this book will help all of us involved in the International Year to bring an end to the last great taboo.

This is the moment, long overdue, to set a new sanitary revolution in motion. We need the words, the courage and the dedicated resources to do what we must to make a difference.

His Royal Highness Prince Willem-Alexander
Chairperson, UN Secretary General's Advisory Board on Water and Sanitation

Preface

More than 1.2 billion people worldwide gained access to improved sanitation between 1990 and 2004. However, even with this progress, some 41 per cent of the world's population – an estimated 2.6 billion people, including 980 million children – lack access to proper sanitation facilities.

Inadequate sanitation, hygiene and water are serious global problems that contribute to the deaths of some 1.5 million children under the age of five annually, largely due to diarrhoeal diseases. Many millions more suffer repeated bouts of illnesses that damage their health and nutritional status, and keep them out of school. In the majority of these cases the underlying problem is the failure to dispose safely of human waste and to prevent pathogenic particles finding their way onto hands and food and into drinking water.

If sanitation is improved, lives will be saved. But the impact of poor sanitation extends beyond health. Where schools do not provide proper toilets for children, and particularly for girls, their educational prospects suffer. Faced with a lack of girl-friendly facilities, many parents withdraw their daughters from school when they reach adolescence.

The evidence tells us that education – especially of girls – is critical for development and for the empowerment of women. It raises economic productivity, reduces poverty, lowers infant and maternal mortality, and helps improve nutritional status and health. Clean, safe and dignified toilet and hand-washing facilities in schools help ensure that girls get the education they need and deserve. When they get that education, the whole community benefits.

Although UNICEF's involvement with programmes to provide basic sanitation facilities to children and their families began more than 30 years ago, and the organization has helped develop and distribute low-cost toilet technology to countries in Asia and southern Africa for many years, much more needs to be done.

The authors of this book, Maggie Black and Ben Fawcett, acknowledge that the issue is a sensitive one for many people. As a result, insufficient global attention has been given to improving toilet facilities, and to the economic and health benefits that would follow. It was for this reason that, when the authors came to us with a proposal for a book that highlights this neglected health and development issue, UNICEF decided to lend its support.

The book examines the history of sanitary reform, and describes the innovative work undertaken by public health engineers and experts all over the world, including those working at the community level whose efforts have been essential to the progress made over recent years. The authors pay particular attention to the impact of sanitation on children's health, and on the role that children can play in improving their circumstances and those of their families and communities. Students who take part in a regular school-cleanliness regime take forward their new knowledge and habits into their lives, insisting on sanitary behaviours in their own future homes. On their shoulders rest the best hope for a domestic sanitary transformation worldwide.

When the UN General Assembly declared 2008 the International Year of Sanitation, it called for increased international attention to address the impact of the lack of sanitation on health, economic and social development, and on the environment. UNICEF hopes this book will play a part in increasing that attention.

Ann M. Veneman
Executive Director, UNICEF

Acknowledgements

This book is a collaboration between a writer on watery affairs – who did the writing; and an engineering practitioner, researcher and lecturer on low-cost sanitation – who undertook the research, technical oversight and quality control. It came out of the engineer's reaction to an earlier book by the writer on the story of UNICEF's contribution to the development of water and sanitation services in India. It was time, he suggested, for a book in a similarly accessible style on the subject of the sanitary crisis in the developing world. When sanitation is treated as an adjunct to water, it typically disappears – or, if examined at all, becomes buried in one chapter among many. Despite the writer's initial reservations that people don't normally want to hear about, read about, or publish material on the disposal of human waste, a project was born.

At that time, and equally at the moment in 2006 that UNICEF committed the kind of support that made the project feasible, there was no expectation that the UN would declare 2008 the 'International Year of Sanitation'. It seems, finally, that the taboos which prevent the subject of excreta being properly aired in debates on poverty and international development are succumbing to the scandal that 2.6 billion people in the world at large are without a decent place in which to perform unavoidable bodily functions. The book has caught an unexpected zeitgeist. If it can help in a small way during the course of sanitation's year in the spotlight to speed up a much-needed sanitary revolution, similar to the one that sewered and cleaned up industrialized environments in the 19th and early 20th centuries, the authors will have achieved what they set out to do.

At first glance, this is undoubtedly a distasteful and difficult subject, especially in societies that are excrementally squeamish – as most of them are. But once the first inhibitions are conquered, everyone with whom the book has been discussed quickly discovers that this is also a fascinating subject. A lot depends – as in all matters where intimate personal behaviour is involved – on how it is handled. This has been a challenge, especially to the writer; public health engineers may have no more trouble discussing U-bends and 'dry toilets', and the composition and removal of their stinking contents, than experts on HIV/AIDS have in discussing sexual organs and acts of copulation. But for most people, that is not the case. Presenting the relevant information about s**t in language and terms that will be acceptable to a wider audience is no easy task. Breaking these taboos will be difficult for many of the dignitaries involved in the International Year, and for the media reporting on it in many countries. But the authors have found that it can be done.

The book begins with the story of the first sanitary revolution, in an attempt to learn from the energy and commitment of earlier sanitary pioneers. The Victorians managed to encompass dirt, filth and its disease-spreading properties in the ongoing discourse of their age – because they had to. So do we, if we are to address the global sanitary problem. The authors discovered that there are unsung heroes of low-cost sanitation – of the kind that would make life so much more pleasant, comfortable and health-protecting for the people of the developing world – in the contemporary era. Several of these, some of whose names appear in these pages, began many years ago to try and 'talk up' sanitation and get those in government and the international community with influence and resources to listen. The late Jim Howard of Oxfam, well known to both authors and key mentor to one, was a brilliant early advocate of the need to address people's excretory requirements, especially in emergencies. Jim's debriefing on the situation in the 'shitting fields' – the desperate famine camps of 1984 Ethiopia – was an inspirational turning-point in the engineer's career. Martin Beyer, chief of UNICEF's Water and Environmental Sanitation department in the 1970s and early 1980s, was another early champion of the idea that sanitation should be neither an ornamental nor a dirty word. Beyer, who launched the writer in this partnership down the path of watery and sanitary prose, therefore also deserves special acknowledgement. So does John Kalbermatten, retired senior policy adviser on water and sanitation at the World Bank, inspirational force and father figure of the 30-year international effort to promote 'sanitation for all'.

A number of other people deserve recognition as pioneers, colleagues, friends and helpers on the sanitary frontier or as toilet missionaries in difficult circumstances. Many have given us assistance during the course of the book's preparation or we have drawn heavily on their existing work. Special thanks go to Deepa Joshi, Gift Manase and Martin Mulenga of the Institute of Irrigation and Development Studies at the University of Southampton, whose insights and output proved invaluable in a ten-year research programme into the social and institutional aspects of a much neglected subject. Others include: Anita

Abraham, Nisha Bakker, Francisco Basili, Lizette Bergers, Björn Brandberg, Sandy Cairncross, Paul Calvert, Renato Chavarría, Manus Coffey, Val Curtis, Rafael Díaz, Paul Edwards, John Eichelsheim, Carmen Garrigos, Ian Hopwood, Fr. Edwin Joseph, Racine Kane, Alok Kumar, Nilda Lambo, Michel Messina, Richard Middleton, Marina Morales, Joy Morgan, Peter Morgan, Sunita Narain, Claudio Osorio, Bindeshwar Pathak, Dina Rakotoharifetra, Michel Saint-Lot, David Satterthwaite, Chandan Sengupta, T. S. Singh, Steven Sugden, Nienke Swagamakers, Rupert Talbot, Kevin Tayler, Henk van Norden, Christine Werner and Uno Winblad. We would especially like to thank Therese Dooley, who has given us unstinting support and feedback from the moment she arrived in UNICEF's headquarters as Sanitation Officer and found this project on her plate. Also Vanessa Tobin, then chief of UNICEF WASH, who backed the project from the start, and her successor, Clarissa Brocklehurst, as well as Nolina Gauthier at UNICEF for her administrative help. We would also like to thank colleagues in UNICEF, WaterAid UK and WEDC, Loughborough University, and Uno Winblad, David Eveleigh, Punch Magazine and the Mary Evans Picture Library for their assistance with the illustrations. Finally, we thank Rob West, Hamish Ironside and Alison Kuznets at Earthscan for taking on the book's publication, and their help during its preparation.

Maggie Black *Ben Fawcett*
Oxford, UK *Mullumbimby, Australia*

List of Acronyms and Abbreviations

AGETIP	Agence d'Exécution des Travaux d'Intérêt Public (Agency for Implementation of Public Works, Senegal)
CAPS	Comité para Agua Potable y Saniamento (Committee for Drinking Water and Sanitation, Nicaragua)
CBO	community-based organization
CIS	Commonwealth of Independent States
CLTS	community-led total sanitation
COW	contractor-oriented work
CRSP	Centrally Sponsored Rural Sanitation Programme (India)
CSE	Centre for Science and Environment (India)
DFID	Department for International Development (UK)
DPHE	Department of Public Health Engineering (Bangladesh)
DSK	Dushtha Shasthya Kendra (a Bangladeshi NGO)
DSSD	Dar-es-Salaam Sewerage and Sanitation Department (Tanzania)
DVC	double-vault composting (latrine/toilet)
ecosan	ecological sanitation
ENACAL	Empresa Nicaragüense de Acueductos y Alcantarilladoes, the state water company in Nicaragua
EU	European Union
FISE	Emergency Social Investment Fund, Nicaragua
FSG	Frères Saint Gabriel (an NGO in Madagascar)
GOI	Government of India
IDWSSD	International Drinking Water Supply and Sanitation Decade (1981–1990)
IMF	International Monetary Fund
KWAHO	Kenyan Water for Health Organization
MAPET	manual pit emptying technology
MDG	Millennium Development Goal

MP	Member of Parliament (UK)
MSPAS	Ministerio de Salud Publico y Asistencia Social (Ministry of Public Health and Social Assistance, El Salvador)
NGO	non-governmental organization
O&M	operation and maintenance
OPP	Orangi Pilot Project (Karachi, Pakistan)
ORT	oral rehydration therapy
PHAST	participatory hygiene and sanitation transformation
PHC	primary health care
PHE	public health engineering
PRRAC	Projecto de Reconstruccion Regionale de America Centrale (Central American Regional Reconstruction Project)
PRSP	poverty reduction strategy paper
PSP	private sector partnership
R&D	research and development
SANA	Sustainable Aid in Africa International (Kenya)
SANAA	Servicio Nacional de Aguas y Alcantarillas (National Autonomous Water and Sewerage Authority, Honduras)
Sida	Swedish International Development Cooperation Agency
SKA	Safai Karamchai Andolan ('sweepers' movement') (India)
SPARC	Society for Promotion of Area Resource Centres (India)
SSHE	school sanitation and hygiene education
SSP	small-scale provider
TAG	Technical Advisory Group (of the World Bank)
TB	tuberculosis
UD	urine diversion
UEBD	Unido Ejecutivo para Barrios en Desarollo (Executive Unit for Developing Settlements, Honduras)
UNDP	United Nations Development Programme
UNICEF	United Nations Children's Fund
USAID	United States Agency for International Development
VBW	village-based worker
VERC	Village Education Resources Centre (Bangladesh)
VIP	ventilated improved pit (latrine/toilet)
W&S	water and sanitation
WASH	water, sanitation and hygiene
WC	water closet
WHO	World Health Organization
WSP	Water and Sanitation Program (originally managed jointly by the UNDP and the World Bank, now a World Bank programme alone)

1

A Short History
of the Unmentionable

Previous page: Peckham sewer: this illustration from *The Illustrated London News* (1861) shows tunnelling for a sewer in London during the construction of the drainage works for the metropolis directed by Sir Joseph Bazalgette.

The disposal of human wastes is a subject normally buried in euphemism and avoidance – at least in public. Privately, every single person on the planet is intimately concerned on a day-to-day, even hour-to-hour, basis with the need to relieve themselves in a congenial place and fashion. The physiological necessity of excretion cannot be averted, even if it cannot be spoken of. No-one can attain the purity of the perfect saint who manages to digest every single thing he or she consumes – a power that certain holy men in the ancient world were believed to possess. Even the most rigid toilet-training can enable a person to postpone their visits to the 'rest-room' only for a certain length of time.

Every day, each human being emits an average of slightly less than 100 grams of faeces and roughly one and a half litres of urine. Even though this regular process of bodily evacuation may not be thought of as 'dishonourable and base' – a phrase from Victorian England – it is nonetheless regarded as noxious and unmentionable by most of the human race.

One of the dominant euphemisms in the business of sanitation is the term used to describe the infrastructure for removing human excreta from our homes, offices and buildings: 'public health engineering'. In the industrialized world, this means pipes bringing clean water in, and other pipes taking the same water away once it has been dirtied and flushed. But the only part we recognize is the water: we pay 'water rates' for 'water connections'. We talk about 'water-related' disease when most disease is sanitation- and hygiene-related. We don't mention the s**t. Yet for millions of people, especially children, whose lives are threatened each year by bouts of diarrhoea, disposing of that is by far the more critical problem. For urine is virtually free of hazardous material. Only the s**t contains large numbers of pathogens dangerous to health.

This is an unspoken subject in almost every culture, except in the context of ribaldry and scatological humour. The organs of defecation are close to or identical with those used for sex – another delicate subject. Those down the centuries who have written about toilet habits have found the associated vulgarity and eroticism difficult to negotiate. Comedians may enjoy the subject, but those of an academic or intellectual bent tend to avoid it, and most ordinary citizens adopt an attitude of 'out of sight, out of mind'. The subspecies of engineers who preoccupy themselves with mire are viewed askance as if they were mildly touched, and in the UK plumbers earn more than other kinds of household fixer because of bravery in the face of ordure. Our attitudes are not surprising. Faeces and urine are extremely distasteful and it is preferable to carry on averting our gaze – and our noses. Except that, as a result, the worldwide sanitary crisis is often wrongly diagnosed and wrongly addressed, when it is noticed at all.

That there is a worldwide sanitary crisis is indisputable. Over 40 per cent of people in the developing world still depend on a bucket, a bush, the banks of a stream, a back street or some other sheltered place for their several daily emissions. Apart from the indignity, especially for women, of this lack of decent facilities, the world is daily becoming more crowded and more urbanized, with all the sanitary complications this represents. The presence of human detritus in many nooks and crannies of the environment – alongside paths, in waterways, in fields, in alleyways – and the unavoidable spread of minute particles onto feet, clothes, hands and faces constitutes a major health hazard, as well as being aesthetically unpleasing to every human sense. According to the United Nations Children's Fund (UNICEF), 1.5 million children under five die each year due to a diarrhoeal disease in which lack of decent toilets and poor hygiene are deeply complicit.[1]

During the last 25 years, there have been a series of international commitments, made at different times in various international gatherings, to the target of 'safe water and sanitation for all'. Every time progress towards the target is reviewed – for example at the end of the Water Decade in 1990, or at the decennial review of the UN Goals for Children in 2001 – the numbers of people who are still without sanitation seem barely to have changed. The latest and most talked-up target is the Millennium Development Goal (MDG): to halve by 2015 the proportion of people who in 1990 were without access to basic sanitation. Although the MDGs were established at a special UN Summit in 2000, the goal on sanitation was a poor relation. It was added to the almost identical goal on safe water as a result of heavy pressure at the second Earth Summit at Johannesburg in 2002. Only after days of intense lobbying did all the nations of the world agree that people everywhere need not only water, but toilets as well.

Since then, sanitation has gradually harvested more international attention – culminating in the designation of 2008 as the 'International Year of Sanitation'. A report on progress towards the MDGs published by the UN in 2006 stated that, during the period between 1990 and 2004, 1.2 billion people 'gained access to improved sanitation' – excluding those whose 'access' is to a public or shared facility whose standards of cleanliness are too often abysmal to qualify as 'improved'.[2] This rate of progress is encouraging, but regrettably it is less impressive than it sounds. Around 2.6 billion people still do not have this 'access', a figure that has not much changed over the years (Figure 1.1). This is because the global pace of toilet take-up barely matches that of population growth in the places that matter. Obviously, people without 'access to sanitation' deposit their excreta somewhere, and many deploy methods which they regard as correct and

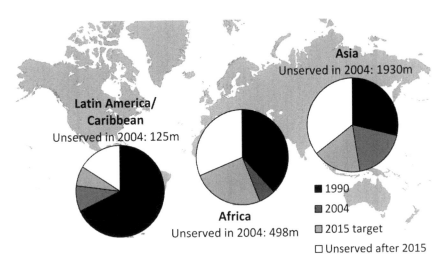

Figure 1.1 *Growth in the percentage of sanitation coverage, 1990–2015*

Source: WHO and UNICEF (2006) *Meeting the MDG Drinking Water and Sanitation Targets: The Urban and Rural Challenge of the Decade,* WHO/UNICEF Joint Monitoring Programme, Geneva and New York

acceptable from a cultural point of view. But whatever these are, they are not regarded as adequate by public health experts, meaning that the s**t in particular is not deposited in a toilet, confined to a sewer or closed container, or rendered safe from contact with domestic water supplies, human touch, food, utensils or other points of contamination.

The very word 'toilet', at least on the Anglo-Saxon tongue, has an awkward and embarrassed ring, and its use is often regarded as vulgar. Minor media skirmishes occasionally break out over the terminology to be used in polite society for the 'convenience'. Whatever the preference of the elect in societies around the world, 'toilet' is the word that has entered international parlance and carries the flag in the formal literature. It is derived from the French 'toilette', meaning the business of dressing, making-up and perfuming oneself in preparation for display in society – and it is salutary to remember the importance of the connection between 'sanitation' and personal grooming. The cultural concept of clean habits and bodily purity has much wider connotations than the scientific idea of safety from disease which dominates the public health sanitary agenda.[3]

The small closet with a porcelain bowl and water-seal U-bend with which the word toilet is now synonymous only entered widespread usage during the European sanitary revolution of the 19th century. This was the period when the template for 'public health engineering' was established, and the flush toilet with its plumbing connections to an adjoining sewer gradually replaced the earth closet, the chamber pot, the privy, the outhouse and other 'dry' systems previously used by the vast majority of people. It also elevated water to the position of supreme sanitary agent, the factor that made it possible to bring about major advances in public health and life expectancy. As a cleansing agent, water cannot be surpassed: it has extraordinary powers of dilution, dissolution and absorption. When it acts as a seal in a U-bend, it blocks out all the bad gases, as well as winged bugs and other creepy-crawlies who might want to make a reappearance up the tube. Water flushes and, under pressure, scours. Its insistence on movement downhill on the easiest of gradients, and convergence with streams and rivers flowing to the sea, provides the world with a natural inbuilt lavatory and washing-up apparatus.

Water's cleansing, deodorizing and health-giving properties have always been revered, in the traditional world by priests and pilgrims and in the modern by agents of cleanliness. These roles are often conflated: priests and spiritual leaders have been frequent movers and shakers for sanitary improvement. Many of the world's rivers are regarded as holy, and taking a bath in them is seen as purifying. On India's *ghats* – flights of steps leading down to the river – praying, bathing, laundry, 'toilette' and recreational swimming all mix together. Unfortunately, the crowdedness of the modern world has overwhelmed the capacity of streams to absorb the necessary volume of dirt. Many of the rivers and lakes where people wash and children romp are filthy. In India, 80 per cent of the pollution load destroying the country's rivers is untreated human waste.[4] In much of the developing world, only a fraction of sewage and drainage water is treated before being discharged into waterways.[5]

When 19th-century sanitary architects and engineers began to construct their tunnels, pump houses, sewers and treatment plants to enhance or replace the self-cleansing powers of the natural environment, they created a new political economy surrounding human wastes. Providing every urban household with a supply of flowing water sufficient for drinking, cooking, washing, laundry and flushing, and also a system to remove the dirty water and sewage, was extremely complicated and expensive – hence the 50 years or so needed to bring it about. Yet so squalid and disease-ridden had towns and cities become in the industrializing world of Europe and North America that the necessary resources were found, as a product of the industrial progress that spawned the problem in the first place. Although the demand for toilets and bathtubs generated good livings for private entrepreneurs and manufacturers, when it became clear that the poorer, more squalid parts of town would never be serviced with functioning human-waste disposal unless the state took a hand, resources from the public purse were provided. In the interests of public health, a duty to ensure that every household had a clean water supply and means of safe disposal of wastes regardless of its ability to pay eventually became widely accepted.

The sanitary revolution of the 19th century brought huge benefits to humankind. The public health engineering model developed in Britain, Europe and North America was exported all over the world. As towns and cities mushroomed, the task of building and managing municipal water supplies and sewers was assigned to publicly funded authorities, in Djakarta, Buenos Aires, Delhi and Nairobi as in Washington, London or Berlin. But the model, for all its achievements, contained several inbuilt flaws.

Principally, conventional water-borne sewerage is not an affordable way of dealing with the sanitary crisis in non-industrialized, low-income communities, and it is impossible to picture a time when it could become affordable in the large parts of the developing world still characterized in this way. These are the environments occupied by the world's most disadvantaged people, including the most vulnerable children and women. The resources are lacking, at both the community and national levels, to provide sewers and flush toilets for the majority of the inhabitants. They are also lacking to lay on water in sufficient quantities and at sufficient pressure for the kind of heavy domestic use standard in industrialized settings. And even if hundreds of billions of dollars could miraculously be found to pay for the construction of pipes and sewers and for the water to flow through them, their proper management and maintenance would be far too expensive, and the pollution they would add to already heavily polluted rivers could lead to further public health disasters.

Another legacy of the sanitary revolution is that the inhabitants of sewered environments have a subconscious tendency to conflate water and human waste disposal in one water-dominated paradigm. They, or we, forget about the inevitable wastes – where they go, who has to deal with them, and the technology required to transport and treat them. This mindset, with its underlying assumption that a water supply is the be-all and end-all of public health, has become extremely unhelpful. It is an important reason why the original inclusion of a Millennium Development Goal for water supplies was taken for granted while one for sanitation had to be hard fought for. Furthermore, the mindset is reflected among those in authority in the countries most affected by hygiene-related diseases. 'Sanitation' is invariably a poor relation to 'water' in the public health engineering portfolio and receives far less resources. Water supplies, for which there is strong demand and considerable political enthusiasm, consistently improve. Meanwhile, personal hygiene and waste-disposal facilities, for which there is less overt and forceful demand, and substantially less political and therefore donor interest, consistently languish.

Most of the 2.6 billion people without access to basic sanitation live in the towns and villages of the developing world and, for them, the means by which most of them dispose of their excreta now, or could dispose of it in the future, is entirely separate from their water supply: there is literally no connection. In a majority of cases, their toilet – if they get to have one – cannot have its water supply piped in and its output piped away: neither they nor their local authorities could afford the several hundred dollars of investment per household required. Something else – a ventilated earth closet, a pit with a toilet flushed by pouring a jugful of water down the pan, access to a neighbourhood amenity, a more modest tank and drain configuration, a plastic bag thrown on the garbage, a scavenging pig or dog, a walk in the fields – has to suffice.

Even in the industrialized world, and certainly in the rapidly industrializing world, where demand for public health infrastructure is growing fastest, the costs of sewerage are exorbitant: it costs hundreds of billions of dollars to meet the sanitary standards laid down by EU or US environmental regulations. In most settings, water is treated and sanitized in such a way that all water piped to households is potable – regardless of the fact that most of it is going to be flushed down some kind of drain without going near a tumbler or cup of tea. These treatment costs are unthinkable in most developing environments. In addition, many of the poorer countries in Africa and Asia are water-short. Supplying households with the 15,000 litres of water per person per year used in Europe or North America to flush away excreta is unfeasible. Even in the industrialized world, if we were starting again from scratch today, many public health enthusiasts would have second thoughts about promoting conventional water-borne sewerage as the one and only sanitation solution, given all the upstream and downstream costs even in water-rich environments. For countries suffering from water stress, and simultaneously from heavily polluted rivers, universal sewerage of the conventional kind is a non-starter.

Nonetheless, in Asian, Middle Eastern and African towns and cities suffering from just such environmental stresses, water-borne sewerage using similar specifications to those in highly industrialized societies is still so promoted. Water closets are the toilet of superior choice throughout the world, and no-one who is anyone wants to endure the humility of inferior domestic arrangements. But since the flush toilet with an outlet to an underground sewer is unaffordable for the vast majority, most people simply go without decent sanitation of any kind.

There is, however, no real excuse for this desperate dearth of facilities, even if 'demand' is not yet powerfully expressed. It is true that some people currently without toilets have reservations about certain types of 'dry' sanitation – especially where such facilities are designated as inferior by calling them 'latrines'. People have a natural conservatism about intimate behaviour, and changes in personal hygiene may require abandoning long-held beliefs about what is clean or unclean. But the fact is that there are affordable alternatives to the sewered flush toilet which are compatible with people's cultural and religious sensitivities; it is simply necessary to make the effort to explore the options and make them respond to the requirements of those in need.

The lack of decent facilities in ever more crowded towns and villages presents many problems, especially for those – mostly women – whose lives are principally confined to activities in and around the household. Where custom and modesty require that women go out only after dark for their daily business, the lengthening distance they have to walk for cover may expose them to danger. In crowded shanty-towns, personal safety when using communal facilities is also a problem. At schools in many African and Asian and some Latin American settings, there may be no facilities, or no separate facilities for boys and girls; and what there are may be foul and poorly maintained. There is strong evidence that the lack of anywhere to relieve themselves in privacy and decency is an important reason why girls are kept out of school, especially after they reach puberty and have also to handle the problems associated with menstruation (such as changing and disposing of cloths).[6]

Tackling today's world sanitary crisis requires another sanitary revolution, and if the international goal of 'sanitation for all' is ever to be met, there is no time to lose. The first requirement is a shift from conventional forms of water-borne sewerage as the one and only solution to human waste disposal, to alternative, cheaper and more sustainable systems, and attracting into them investment and effort from public and private sources. This requires the creation of a new political economy surrounding human wastes. It also requires developing a better understanding of existing hygiene habits and consumer demands among communities which will never be able to have sewers and which appear up to now to have expressed little enthusiasm for the installation of toilets of any kind.

The new sanitary revolution will also require some de-linking of water supplies from sanitation in the public and official mind, so that their respective roles in the causes and prevention of illness are more carefully defined and better understood. It will also mean finding ways to build real political commitment behind sanitation, both in the local government and national institutions of the developing world and in the international donor community. And that in turn requires that the squeamishness that surrounds the subject with silence and taboo is tackled head on. In the same way that the global epidemic of HIV and AIDS has brought us to talk about hidden forms of sexual behaviour, today's sanitary crisis requires that we dismantle the last great taboo, and learn to talk about … shit.

The history of sanitation is as old as the history of human settlement, although the availability of information on the subject is patchy and haphazard. Wherever people congregate and live in close proximity, some organized system of depositing and removing their wastes becomes necessary. Archaeologists have uncovered the remains of sanitation networks and even the odd flush toilet from some of the very earliest civilizations. The Harappans who lived in Mohenjo-daro in the Indus basin in 2500 BC, for example, had a highly developed system made of brick, in which wastes from each house flowed into an adjacent drain. Excavations in cities in Mesopotamia from a similar date have also exposed brick sewers, with lateral drains connected to water-flushed toilets. The palaces of the Minoan civilization in Crete, dating from 3000 to 1000 BC, contained terracotta pipes which carried water under pressure to fountains. They also had bathtubs not unlike our own and elaborate stone drains which carried sewage, roof water and liquid wastes.[7]

The Romans were famous for their sanitary activities, and even had gods of ordure and toilets – Stercutius (Saturn) and Crepitus respectively – and a goddess – Cloacina – of the sewer or *cloaca*. But their sewers were primarily conduits for surface drainage – rainwater and its accumulations – and did not receive human excrement, which was thrown into the streets. Here it lay in a runnel down the middle until the street was flushed with water. Frontinus, the Roman water commissioner in AD100, complained that so much water was diverted from the city's aqueducts for settlements round the city that it was impossible to conduct street cleaning. Most Roman gentry used chamber pots, which were emptied by slaves, while the common people used the public facilities and bath-houses. The emperor Heligogabalus, described as 'a monster of lust, luxury and extravagance', owned 'close-stool pans of gold, but his chamber pots were made some of myrrh and stones of onyx'.[8]

Monasteries, with their concentrated populations, also had to cope with the business of human wastes. The great Buddhist city of Anuradhapura, the seat of Sri Lankan kings for a millennium from around 400 BC contained monasteries housing thousands of monks. Their facilities were elaborate: stacked and porous urine-pots were designed to filter their contents, leaving pure water in the lowest container. Excrement was used, as in many agriculturally pressured civilizations until today, for fertilizing the fields: in China, over 90 per cent of human excreta is still used as manure.[9] In London, up until the sanitary revolution of the 19th century, gangs of nightmen were employed during the hours of darkness to dig up the foul-smelling contents of cesspools in people's backyards. These were carried through the house into carts and barges. Once dried out in holding areas, the residues were ferried up the Thames for onward passage to the countryside in Hertfordshire or Hampshire.[10]

Castles and forts in mediaeval Europe had *garderobes* which emptied down chutes or down the outside of the building, into the moat or onto the rocks below. The *dzongs* of Bhutan – which doubled as monasteries and military forts – had wooden chambers jutting out from their sides, with floors of open joists over whose gaps monks and soldiers squatted. In India, castles and forts had protrusions over rivers or open ground. The river bank or sea shore was the accepted site of 'open defecation' in the Indian sub-continent, and for many people today remains preferable to seclusion in a dark, and usually smelly, little shack. In the city of Sana'a, Yemen, the toilet was a tiny room at the top of the house with a long drop to street level. Dry faeces were periodically taken away via hatches in the street, to be further desiccated in the sun and used as fuel.[11] In Kabul, Afghanistan, and in other towns of Central Asia, traditional dry vaults were similarly emptied and taken by donkey cart to be used for agricultural purposes outside the town.[12] At least in these cases the material remained confined during dehydration. Most arrangements in mediaeval European cities were far cruder, with excreta freely jettisoned into the streets. In Berlin the refuse heaps piled so high in front of St Peter's Church that, in 1671, a law was passed requiring every peasant who came to town to remove a load when he returned home.[13] In Denmark, cleaning the public latrines was the job of the hangman. The removal of human dirt has carried stigma in every society, even if in less crushing a form than that borne by the sweepers of India, designated as untouchable because of their occupation in life.

Paris in the late Middle Ages and early modern times was the metropolis of Europe and the byword for refinement and fashion. But its streets stank from the languishing contents of chamber pots hurled from the windows, a procedure which also posed a risk to passers-by. The palaces of the Louvre, the Tuileries, Versailles and Saint Germain, where aristocrats mingled to curry favour with the King, have ever since been notorious for the nuisances deposited indiscriminately by courtiers behind curtains and doors, on staircases, balconies and in courtyards, without any attempt at disguise. Their lavatorial behaviour was no model for ordinary people, who did likewise in the streets. However, a Frenchman visiting an English country house in the 18th century confessed himself disgusted by English habits, notably the way that 'the sideboard is garnished with chamber-pots' which people regularly got up from the table and went over to use, in full view of others who were still at table drinking.[14]

Mediaeval towns and cities in England were filthy. Although records exist of public privies situated over streams, their provision was grossly inadequate. Most privies, public and private, disgorged directly into cesspools or 'middens', often directly underneath the sitting-place. Many of these constructions were permeable, and the liquid leached away to be absorbed by the soil. Others were of solid masonry, and though in palaces and castles they might connect somewhat haphazardly to drains, most of them had to be emptied by 'rakers'. These were well paid for their sordid work, and they also sold on their harvest of sewage to farmers whose fields lay close to the city walls. Later, another market emerged, for the nitrogen content, among saltpetre men making gunpowder for the Spanish wars.[15] Occasional accidents occurred. In 1326, the Coroner's Roll recorded that Richard the Raker fell through a rotten floor into a cesspit and drowned in its contents. Such stories were not uncommon and are repeated with relish in contemporary accounts of indignity.

Where 'necessary rooms' were installed in town houses, they were often in the basement and prone to overflowing. 'Going down to my cellar,' wrote Samuel Pepys in October 1660, 'I put my foot into a great heap of turds, by which I find that Mr Turner's house of office is full and comes into mine.' The stink from below often fouled the air throughout the upper floors. In country houses, such a room could be placed on the ground at the far end of the building, making emptying less problematic. No wonder that the better-off preferred their chamber pots and 'closed stools' – portable privies with seats of padded leather or, for royalty, red velvet – whose contents were dealt with by the servants. Different receptacles for different excretory functions were common, at least in polite society. In the countryside, the outhouse with its bench of sitting holes was the standard family resort. But disposal in urban areas remained the problem. In London, the discharge of any water except kitchen slops into the drains was prohibited by law until 1815. In Paris, the prefecture postponed until 1852 an injunction that all new and renovated buildings should discharge waste to the sewers – leading to an extensive sewer construction programme.[16]

If European civic administration regarding waste disposal was inferior to that of several ancient civilizations, improvements on the 'closed stool' were more forthcoming. In 1592, a water closet designed by Sir John Harington, a godson of Queen Elizabeth I, made a well-publicized debut. He installed a gadget of his own design in his house at Kelston near Bath, where he received a visit from Her Majesty. It seems she was impressed, as one was later installed in Richmond Palace. Although neither example survives, Harington did write a book: *A New Discourse on a Stale Subject: Called the Metamorphosis of Ajax*. 'Ajax' was a pun on the word 'jakes', contemporary slang for privy or closet, and the treatise was light-hearted, describing how to transform 'your worst privy' so that it was 'as sweet as your best chamber' all for a mere 30 shillings and eight pence (see Figure 1.2).[17]

Although Harington hoped to grow famous on the basis of his invention, it was not widely taken up, although a few travellers and diarists of 17th-century England do mention encounters with 'pretty machines in the House of Office' using water to flush away excreta. By the 18th century such machines had made their way to France, where an architectural work of 1738 included designs for what were euphemistically referred to as *'lieux à l'Anglaise'*. Gradually they became better known, but the first English patent on a water closet was taken out only in 1775, by a London clockmaker called Alexander Cummings.

Thus the protracted development of the water-flush toilet was inspired by the need to provide superior conditions of easement for those at the pinnacle of society: public health considerations played no role at all. For a queen or a duke,

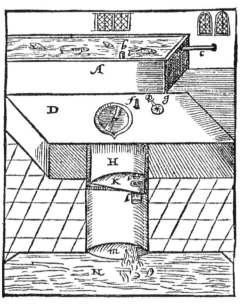

Figure 1.2 *Harington's water closet: Fit for a queen*

Sir John Harington's improved 'jakes' of 1592 (see text). The water cistern is A, the seat D, the stool pot H; water was released from the cistern to a point just below the seat, and the key (g) turned to open the brass sluice (K) and flush the waste into the water below at N.

Source: David J. Eveleigh (2002) *Bogs, Baths and Basins: The Story of Domestic Sanitation*, Sutton Publishing, Stroud, UK

and thus for a squire and his lady, the 'necessary' must be accomplished in comfort, cleanliness and convenience, without lingering malodour. Cummings' key innovation was to introduce a water-seal trap in the shape of a U, in which the water was completely replaced at every emptying of the pan, making the device fully self-cleansing. This meant that matter could not remain in the pipe below, and gases could not re-enter the room from underneath. The curious feature of Cummings' water closet – a feature common to 'improved' WCs for some time to come – was that its design, with elaborate inlet and outlet valves, was much more complicated, and more inclined to fouling or faulty function, than the simpler flushing toilet in use today. This was invented in the early 19th century as a pared-down device for the inferior sort of customer unable to afford 'the best'. The 'cottage pan' with its simple basin and trap began to be promoted by sanitary reformers in the 1840s as ideal for use by the poor.

The flurry of innovatory activity to do with water closets and patents that took place in late 18th- and early 19th-century Britain belonged to a much broader commercial drive based on a new market for home improvements. The early 1800s saw a boom in house building for middle-class inhabitants of towns and cities. Standards of living were rising, along with incomes and expectations of domestic comfort. Manufacturers responded by turning out all sorts of items for the new class of consumers – earthenware, ironware, lamps, fireplaces, cooking stoves, as well as sanitary ware. The boom was to continue throughout the 19th century, making modest fortunes for such entrepreneurs as Thomas Crapper, a metalworker whose name became synonymous with the devices he promoted.

Improvements in bathrooms and water closets, and their take-up on a scale well beyond the most privileged members of society, were therefore an integral part of the transformation in living habits and domestic behaviour which accompanied the British industrial revolution. The terraces, squares and crescents that sprang up in Bath, Brighton, Bristol and Tunbridge Wells during the 1820s were almost certainly the first housing projects anywhere in the world to include water closets as a standard item: they appeared in advertisements for 'gentlemanly residences'. By the 1840s, in houses of a certain standing, the water closet had entered general use. After this, the possibility of separate disposal of the contents of 'stools' and chamber pots disappeared: everything was flushed. And since the development of this household item ran so far ahead of the development of sewers, a major problem arose. The contents of the water closet continued to end up in cesspools situated in the garden or backyard. With the increase in fluid volume, these overflowed more regularly than they had in the past. In 1810,

London, with a population of more than 1 million, was thought to contain some 200,000 cesspools.[18]

Edwin Chadwick, the pre-eminent campaigner for sanitary reform and author of the 1842 *Report on the Sanitary Condition of the Labouring Population of Great Britain*, drew attention to the filthy consequences of these pits, whose contents working people could not afford to empty. Nightmen had found the market for their product dwindling as London expanded. Farms were further away, and they had to charge more for the muck because of the longer distance that it had to be carted or ferried. In 1847, following the arrival of cheaper and more manageable guano – solidified bird droppings – from South America, the market for human waste as agricultural fertilizer collapsed. So people were forced to empty liquid sewage into the street and trust it would find its way into a drain, and from there into a stream or river. Accordingly, the water quality in the rivers running through every town rapidly deteriorated. This happened most famously in London during 1858.

The Thames had gradually come under increasing pressure, both from the demand on its supplies from the rapidly growing population and from the growing volume of excreta finding its way into drains and poorly constructed sewers. A famous builder, Thomas Cubitt, observed in 1840 that instead of every house having a large cesspool as had been the case 50 years ago, 'the Thames is now made a great cesspool instead'.[19] Efforts to achieve sanitary reform and a total reconstruction and systematization of the London sewers had been under-way since Chadwick's 1842 report, but the necessary agreements and technical imprimaturs remained endlessly bogged down in disputes about costs, outflow sites and the administrative responsibilities of different municipal boards.

A long, hot summer in 1858 reduced the Thames to a scandalous condition known as the 'Great Stink'. For weeks, this was the subject of morbid commentary and satirical lampoon throughout the London press. The smell coming off the river was so excruciating that Parliament could barely sit, and sessions in the adjoining Courts of Law had frequently to be curtailed. London suffered regularly from cholera epidemics, and it was still almost universally assumed that air-borne 'miasma' was responsible for its spread. The foulness in the air was therefore regarded not only as horrid, but as pestilential. For many weeks, blinds saturated with chloride of lime and other disinfectants were suspended before every window in the Houses of Parliament. The Thames fishermen had already lost their trade some 20 years before when the last salmon had been landed. Now river boatmen lost their custom and travellers made huge circuits to avoid having to breathe in the fumes (Figure 1.3).

Figure 1.3 *The Silent Highwayman*

Death stalks the Thames: Cartoon from *Punch* magazine during the 'Great Stink' of July 1858

Source: Punch Ltd (www.punch.co.uk)

The Great Stink went on for several weeks, and its supposed threat to life had a powerfully concentrating effect on MPs' legislative capacities. On 2 August 1858, the House of Commons passed an act 'to extend the powers of the Metropolitan Board of Works for the purification of the Thames and the Main Drainage of the Metropolis'. This was the final spur to the transformation of sewerage in London by Sir Joseph Bazalgette, and ultimately led to a triumphant public health engineering revolution not only in Britain but throughout the industrializing world. There was a hidden irony in this. According to Dr William Budd, an expert on typhoid fever writing in 1873, when the returns of sickness and mortality were compiled at the end of the summer of 1858, the result showed 'as the leading peculiarity of the season, a remarkable diminution in the prevalence of fever, diarrhoea and other forms of disease commonly ascribed to putrid emana-tions.'[20] The panic induced by stench had actually been misplaced. But by then the die was cast.

The story of the 19th-century sanitary revolution in Britain, and to a lesser extent in mainland Europe and North America, has been retold so often that its main figures have developed a mythological status, and some of its most instructive features for the business of sanitary transformation in the modern era are buried below layers of historical spin. One such feature is the length of time it took. The transformation of the urban living environment into something piped and sewered, with plentiful safe water on tap, not only in the houses of the better-off, with their valve-outlet WCs, but in the cottages and tenements of ordinary working people, took well over six decades to accomplish. Moreover, while this transformation of industrialized urban settings was ultimately credited with eliminating squalor and epidemic disease, in fact the public health impacts – in terms of radically improved life expectancy and infant mortality rates – did not begin to show up until the final decades of the 19th century and were not significant until past the turn of the 20th century.[21]

The long process of legal, municipal and sanitary reform in Victorian Britain was accompanied by heroic struggles by engineers and reformers on many fronts, and many U-turns in public policy. Many original diagnoses of urban public health problems were wrong, or where they were right took time to gain traction. Social and class attitudes about the labouring poor, both urban and rural, were also in the process of transformation, as were all aspects of economic and political life. Industrialization represented an extraordinary social upheaval, of which the sanitary revolution was both a symptom and a result. The struggle to clean up the towns was long and hard, and the much-celebrated legacy of the sanitary component has shaped theory and practice surrounding public health ever since. However, as often tends to happen, during the subsequent export of these ideas and models, including in the imperial era to overseas colonies and to other Asian, Antipodean and American outcrops of metropolitan influence, some of the most important lessons of the process became obscured.

The roles played by the private and public sectors in 19th-century sanitary transformation are highly instructive – and have recently been conspicuously ignored. As the early part of the story has shown, the private manufacturing sector was critical in producing the toilet, along with taps, pipes, pans, basins, cisterns, U-bends, valves, cocks, spigots, and all kinds of bathroom, hygiene and sanitary ware. All this happened in response to demand for home improvement. However, the mass disposal side was another matter. To begin with, private companies were much involved in water supply and sewerage construction – no other providers were envisaged, even by reformers such as Edwin Chadwick. But at the same time, the leaders of the sanitary movement were convinced that the extraordinary state of filth in the slums could not be addressed without decisive public action. The roles of local and central authorities became a battle-ground, opening up the idea of political intervention in intimate areas of people's lives. It also became clear that private water and sewerage companies were not willing to provide functioning waste disposal or mains water connections to those outside the 'respectable' classes: the costs were too high and the demand – in terms of ability or willingness to pay – much too low. Surely there are lessons to be drawn from this that are valid for the contemporary world.

Another outcome of the 19th-century sanitary revolution is that the retro-spective benefits in terms of public health have been etched in the universal mind as the primary motivation for sanitary improvements: indeed, the whole discipline of 'public health' was the invention of Chadwick and his allies. Yet 'public health' was a *public good* motivation for change, not a private consumer or market-based motivation. Private consumers, where they had the income, wanted to pee and shit in a respectable, clean and comfortable environment. Those who were poor, whatever their desires, did not have the means to pay for flushing closets, water rates, and nightmen to clean out their cesspools and middens. And their better-off neighbours were in no mind to vote the money to provide the poor with these facilities. The public health motivation only applied when it became clear to the better-off that they were themselves threatened by diseases circulating in the poorer parts of town.

In this, the sanitary reformers had a particular disease on their side: cholera. Cholera struck fear and panic in the urban citizens of Europe from the 1830s to the end of the century. In earlier times, the 'black death', or plague, was the disease curse of crowded urban spaces. Although epidemics of plague still threat-ened Asia, these had effectively died out in the Western world by the late 17th century. But in the early 19th century, cholera arrived from Asia and became the new epidemic killer disease of urbanizing Europe. It first appeared in Britain in 1831, and there were devastating epidemics over the next few decades – notably in 1832, 1848, 1849, 1853 and 1854 – during which thousands of people, not only poor people but others from across the whole social spectrum, fell sick and died.[22] New York suffered its first epidemic in 1832, with a death toll of over 3000 – one in fifty of the city's residents.[23]

Cholera spread with deadly speed and was often fatal in hours. Its nature was the subject of wild speculation among members of the medical profession, and for a long time its cause was obstinately explained by the 'miasma' theory of disease. This theory had been entrenched for centuries – witness the derivation of the word 'malaria', which came from the Mediaeval Italian for 'bad airs' (*mala aria*). Edwin Chadwick resolutely believed that 'all smell is disease', and one of his close associates, Dr Niall Arnott, echoed him in describing the cause of many diseases as 'the poison of atmospheric impurity'.[24] Their enthusiasm for the removal of cesspools and middens was therefore related both to the noxious presence of the dirt and, more significantly, to the disease-spreading nature of the stink. Interestingly, modern research suggests that there is indeed a strong correlation between the instinctive human reaction of disgust and proximity to disease-carrying agents.[25] However, Chadwick and virtually all his reformer contemporaries thoroughly misread the nature of the connection.

The association between cholera spread and foul water was first made in one of the most famous incidents of sanitary history, when Dr John Snow, a pioneering anaesthetist, carried out an epidemiological survey into the extremely high incidence of cholera in a part of Soho, London, during the 1854 epidemic. Snow painstakingly enumerated every facet of the local houses, inns and shops, and the water-consumption patterns of their permanent and temporary inhabitants – a scientific method which was itself relatively novel. He demonstrated that the imbibing of water, or beverages made from water, from a particular public pump in Broad Street was the essential common denominator in the majority of cases. He noted that many people drew water from this pump because they preferred it to that from other pumps; this was the cause of cases outside its immediate vicinity. Having completed his inquiry, Snow went to see the Board of Guardians for the parish – the local council of its time – who ordered the handle of the pump removed.[26]

Recounted with gusto by historians down the years, the closing of the Broad Street pump has become an iconic moment in the birth of public health. At the time, Snow was ignored. The miasma theory was so well entrenched and its supporters such powerful figures that only after another epidemic in 1866 was Snow's evidence of water-borne infection given belated recognition. It took until 1883 for Robert Koch, a German bacteriologist, to identify the cholera bacillus in India and show that it was conveyed in water polluted by the faeces of victims.

Today it is difficult for many of us to evoke a world in which scientific information on a matter of such importance took so long to become established and widely known. Nevertheless, that is still the situation today in parts of Africa, and indeed wherever illiteracy is common, where belief in the miraculous propagation of disease by witchcraft or curse remains current, even among some highly educated members of society. And drinking water preference is still an important consideration in understanding behaviour and convictions in many developing country settings where the supply of water is not standardized or uniformly treated. In many environments where the local water supply is from a natural source such as a well, stream, spring or borehole, taste and temperature are still the principal reasons why people may choose to draw water from one source rather than another – as they were in 19th-century London.

In the subsequent telling and retelling of the glorious Snow moment, a curious transposition has occurred. The lesson of the Broad Street pump passed down to posterity is far more closely associated with the quality or safety of drinking water as the key to disease control than with the dangers of inadequate sanitation. And the pre-eminence of Snow in the story has ejected another important claimant from his share of diagnostic fame.

In the London of that time, the flushing of wastes by water from the large number of water closets recently installed had contributed not only to overflowing cesspools, but to the saturation of the surrounding soil by seepage and leakage of excreta through the porous sides of the pits. Some people, by deepening their pits to receive extra amounts of matter, had sited them in the strata through which flowed the fresh underground water to which street pumps and private wells were connected. Contamination of the water supply was the result. And the same thing was happening in every major city. The mortality rates in some towns in Britain at this time were extraordinarily high – in Liverpool, for example, average life expectancy for an unemployed labourer was 15 years and for the well-to-do only 35.[27] The average age of death in the boom town of Dudley was said to be 16 years 7 months, with around 50 per cent of inhabitants dying before the age of 20.[28] A principal reason for this was that the majority of people drank dilute sewage on a daily basis. But since disease was assumed to fly through the air, the dangers of soil and water pollution were not appreciated and the nuisance of the stinking cesspools continued to be misdiagnosed.

While Snow is the lionized hero of the Broad Street story, there was another, more obscure player, who concerned himself not with the water supply but with the shit. The Reverend Whitehead, curate of a nearby parish, and like Snow a member of the Cholera Inquiry Committee, also carried out a house-to-house investigation in the area, confining his inquiry to Broad Street itself. Both Snow's and Whitehead's reports showed an explosion of fatal attacks on just two particular days, with an immediate decline – which, interestingly, began some days before the pump was disconnected.[29]

Whitehead delved deeper than Snow into the mystery of how and why the well had become infected. At No 40 Broad Street, Whitehead discovered that there had been an earlier case of a cholera-like disease, and that 'dejecta' from this patient had been thrown into a cesspool very close to the well. A surveyor was then called in. He found the brickwork of drain and cesspool highly defective, with a steady percolation of fluid matter from the privy into the well. Whitehead thus not only confirmed Snow's water-borne disease theory, but pinpointed the cause. He also concluded that the water had only been infected for a very few days and that, instead of multiplying, the cholera germs had died out. To this he attributed the coldness of the water – cited by many consumers as the reason they preferred the water from that particular pump. Thus taste and preference may not be so misleading as a disease protection quality as is often assumed.

Although safe drinking water took on an overstated role in much later efforts to address 'water-borne' diseases in the developing world, the Victorians themselves were very focused on sanitation and sewerage. Indeed, their efforts to impose cleanliness on the labouring poor had all the characteristics of a moral crusade. Industrialized poverty on such a scale was a new phenomenon, and the 'barnyard conditions amid stench and filth' which characterized the crowded tenements and alleyways in which poverty-stricken working people lived appalled many contemporary observers, who perceived these unfortunates as a race apart.[30] Similar attitudes persist in some societies today.

As the better-off became bathroomed, toileted and sewered, the British classes became strongly differentiated by smell and look. One sanitary historian of the period comments that the 'great division between the respectable and unrespectable was where and how one relieved oneself', and by implication where parents taught their children to do so.[31] Many doctors kept the door open when poorer patients sought their services and did not allow them to sit in upholstered chairs. Magistrates adjourned their courts and continued hearings in the backyard. A sympathetic slum doctor wrote that he had to 'rouse up all the strength of my previous reasonings and convictions, in order to convince myself that these were really fellow-beings'.[32] Whatever the attitude of the poor, the fact was that, without sufficient water for washing and laundry, toilets or drains, they had no means of surmounting the dirt. Where else could they relieve themselves except in local alleys and courts? In some working-class budgets, as much was spent on soap and washing materials as on fuel, milk or tea.[33]

Edwin Chadwick, whose name above all others became synonymous with the evolution of public health, had no training in medicine or sanitary engineering. He was a lawyer who rose to prominence as the main architect and enforcer of the poor law of 1834. He was much hated for its key principle, which was to make it so difficult to seek public relief that few would try to do so.[34] His search for the causes of indigence – to reduce costs still further – led him down the track of infectious disease associated with filth. His 1842 report into the insanitariness of the urban poor suggested that the mire in which they languished had social as well as biological consequences. It induced a psychological degradation, which could cause desperate people to resort to the gin bottle, or worse – to revolution. In the 1840s, revolution was another part of the miasma afflicting towns and cities all over the European continent. In Chadwick's view, the key to a happy, healthy and docile proletariat lay in sewers and water supplies.

Chadwick was also an enthusiast for the reform of institutions and public administration. When in 1848, thanks to his tireless campaigning, a Public Health Act was passed, the General Board of Health was established with the purpose of forcing British towns and cities into sanitary action. However, the board was denounced as a despotic interference in local liberties. The citizen members drawn from the landed gentry and other leading local families, who typically ran the instruments of local administration, wanted cheap government with low spending, and that did not include extending drains and sewers to the incorrigibly dirty poor.

This situation has resonance with some towns and cities in today's developing world. Power within local authorities may similarly reside in the hands of those whose main idea is to favour their own property, business or commercial interests and whose sense of civic responsibility does not extend to provision of quality basic services to the urban poor. This attitude has been noted by observers of the urban scene in India[35] and of poor urban populations in Central America and elsewhere. Thus in both historical and contemporary experience, civic and social attitudes can have the effect of inhibiting sanitary action in the poorer quarters of town. What has markedly changed is the relative absence today of epidemics of sanitation-related disease, or, where they do occur, the sense of threat to their own health felt by the 'better class of person'. Medical advance as a means of self-protection or containment has come to their rescue.

By the 1860s, municipal attitudes all over the industrializing world were changing. This was the era of political reform in Britain, and the enfranchisement of working people also played a part. Leading industrialists had become convinced of the value of cleaner and more efficient cities, and conditions of housing and environmental filth that had long been fatalistically accepted began to be seen as intolerable. The Sanitary Movement, led by churchmen and philanthropists as well as engineering enthusiasts such as Sir Joseph

Bazalgette, began to harvest results. Not only in London but in Birmingham, Liverpool, Newcastle and elsewhere, the existence of crowded tenements with liquid filth oozing through their walls, with one privy and cesspool shared by hundreds of people, became recognized as hazards for the entire urban population. The campaign enlisted the literary and reforming zeal of novelists such as Charles Dickens, whom Chadwick himself consulted over his descriptions of living conditions in the great towns.[36]

The building of new sewerage systems in the second half of the century represented a massive feat of Victorian civil engineering and showed a new energy and direction in urban planning. But the removal of excreta by underground flow was not greeted enthusiastically by those concerned with the increased pollution of rivers. During the 1860s, the notion that this was also a waste of material which could be used for fertilization of the land began to take hold. In 1861, a professor at Giessen University, Justus von Liebig, published a book entitled *Agricultural Chemistry*, in which he proclaimed, 'The introduction of water closets into most parts of England results in the loss annually of the materials capable of producing food for three and a half million people.'[37] The pollution potential of untreated sewage in waterways, the wastage of water for its transport and the squandering of nutrients valuable to farming also comprise the case put up today in favour of 'ecological sanitation'. The methods for dealing with human detritus and its recycling are not dissimilar either: earth or composting toilets and separation of urine from faeces. Similar, too, are the evangelical credentials of their proponents – in Victorian Britain, the exemplar was the Reverend Henry Moule.[38]

Advances in agricultural science had stimulated both the manufacture of superphosphates – the first chemical fertilizer – in 1842 and the import of guano from Latin America from around the same time. These were expensive, so there ought to be a demand for alternative sources. Von Liebig was not the only person to advance the idea that the vast quantities of human sewage generated by urban populations could manure the fields. In 1860, the Reverend Moule took out a patent on his first earth closet, and within three years James White and Co of Dorchester was manufacturing two of Moule's models. They used dry and sifted earth to absorb 'excrementitious and other offensive matter' to make manure, breaking down the faeces by the action of naturally occurring bacteria in the soil. Moule's toilets were only for shit; urine had to be separately disposed of. As with today's eco-toilets, a door in the back enabled the contents to be removed. In advance of sitting down, soil was tipped into the pan. After use this was dropped either into a bucket (the cheaper version) or into a trough (double the

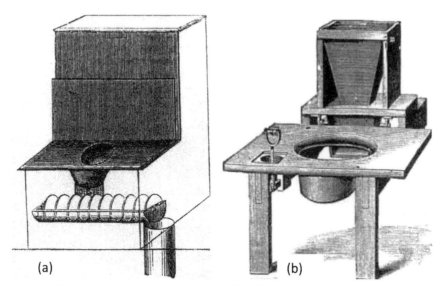

Figure 1.4 *Earth closets designed by the Reverend Henry Moule*

(a) One of the more fancy toilets designed by the Reverend Henry Moule, manufactured in 1863. The contents of the fixed pan beneath the seat led into a trough, where they were mixed together by a rotating screw and deposited down a chute. This model remained on sale until the 1930s.

(b) A later design of 1873 was much simpler: the brass handle was pulled up to draw the hopper forward and release soil into a bucket under the seat-hole.

Source: David J. Eveleigh (2002) *Bogs, Baths and Basins: The Story of Domestic Sanitation*, Sutton Publishing, Stroud, UK

price), where a rotary screw thoroughly mixed up the contents, including cleaning material such as waste paper (Figure 1.4). Earth – both dry and used – was to be stored in a shed. Moule estimated that one cartload would last two or three people from six to twelve months.[39]

Moule was an energetic promoter of his closets, and manufacturers began to produce and market his inventions. They were widely used in prisons and workhouses and also taken up on country estates. Other enthusiasts developed dry privies which employed the ash from household stoves and fires as a deodorizer and drying agent. During the last quarter of the 19th century, upwards of 100 large towns and cities launched schemes for the collection and distribution of sewage as manure on the expectation of healthy profits. Many midland and northern towns, including Birmingham, Rochdale and Nottingham, used a container system. Specially designed pails were given out to householders for regular collection and replacement. The most sophisticated mass collection system, developed in France, used a specially manufactured pail packed with

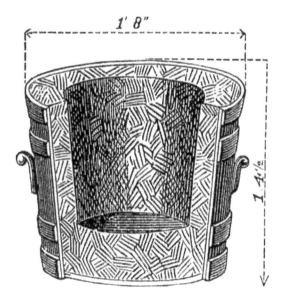

1' 8"

1.4½"

Figure 1.5 *The 'Absorbent Pail' system*

The toilet pail for systematic delivery and collection designed by a Frenchman, Pierre Goux. After emptying, it was repacked with layers of absorbent material 3 inches at the side and 4 inches at the base. The system was used in northern British towns and military camps in the 1870s.

Source: David J. Eveleigh (2002) *Bogs, Baths and Basins: The Story of Domestic Sanitation*, Sutton Publishing, Stroud, UK

absorbent linings made of chaff or straw, and a deodorizer such as soot.[40] The pails were packed with their linings by the operators using a patent mould (Figure 1.5). This system had to be very well managed – the pails thoroughly lined and emptied regularly, and the users had to restrict themselves to faecal matter only.

In most urban and rural areas, the pail system was more basic, and not unlike the use of bucket latrines common in households in the Indian sub-continent until the present day. In Britain, the night-soil men who collected their contents were described as 'very filthy in their appearance and habits', but they did not suffer the indignities of India's 'scavengers' – people whose lot was inescapable. Britain's 'scavengers' – also seen by members of superior classes as remote from any 'tolerable human type'[41] – were people who scavenged for saleable items among the midden heaps and in the gutters. Their equivalent in today's Asian cities are known as rag-pickers, and, like their European predecessors, they are mostly children.

Improvements in environmental sanitation began in earnest in most large conurbations during the 1860s and 1870s. The first stage was the drainage of cesspools, making them smaller, water-tight and thus self-contained. The second step was to introduce dry systems for the poorer urban quarters with organized collection. It took time to move to the third, water-borne phase. Sewers, with their volumes of fluid pollution, were still seen by some as a dangerous and costly fad. While John Ruskin, the artistic and social critic, might rhetorically declare that 'a good sewer' is 'far nobler and a far holier thing ... than the most admired Madonna ever painted', others took a different line.[42] Sewers were still frequently seen as sources of contagion – and with some justice. Outbreaks of typhoid fever – the disease that carried off Queen Victoria's consort, Prince Albert, in 1861 – were traced to defective drains and sewers, and many towns continued to ban the connection of households to drains intended to carry stormwater only.

Only when water was laid on to every house was it possible to begin to replace dry systems with the flushing water closet. In most towns this step did not occur until the 1880s and 1890s. Until flushing toilets could be trusted to work, volumes of water in the mains were adequate, and the technology of sewers and treatment plants had been fully developed and properly installed, the 'dry conservancy' system was less of a risk to health, less foul, less likely to overflow and caused less pollution of rivers. There was also the major question of costs: sanitary engineering was extraordinarily expensive, and financing as well as technological, governance and legal issues dogged urban projects. In most major towns, middens and cesspools were still in use up until the end of the century, although their numbers were gradually declining. As late as 1911, two-thirds of Manchester's inhabitants lived in houses which depended on pails, ash-boxes or a privy midden. In Dublin in the 1880s, 110 nightmen and 39 horse-carts were employed to remove the contents of ash-pits, and Glasgow had 240 'wheelers' on its books, as well as 175 horses and 600 railway wagons, to remove 700 tons of refuse from the city each day.[43]

There were many problems with the recycling of urban excrement as manure. One was the nature of the work, another the question of where to store the muck until it was fit for use. Cartage was expensive, and storage posed problems of public nuisance. Another problem was that, to make the system hygienic, pails had to be sanitized with chemicals such as sulphuric acid, and this reduced the value of the content as a fertilizer. Yet another was the competition from other types of agricultural manure. In 1878, the Local Government Board conducted an extensive survey into town sewage systems and discovered that none of the large towns it inspected had managed to break even on sales to farmers.[44]

Eventually, the use of water for sanitary disposal won the day. Unquestionably, where large and congested municipal populations were concerned, emptying pails and finding places to store and manage the output of 'dry' systems presented public health problems which properly managed sewers did not. The long experience in British towns with 'dry conservancy' has been forgotten, and the lack of profitability and other characteristics which made it inferior to water-borne sanitation, and finally ended its use altogether, ought to be studied carefully by today's enthusiasts for ecological sanitation. The lessons of its abandonment do not mean that improved methods of dry sanitation and nutrient recycling are universally unworkable – the political economy of sanitation in the many different settings of the contemporary world has important differences from those in late 19th-century Europe – but nonetheless valuable lessons may be found.

What cannot be disputed is that, with all the trials and tribulations of its slow adoption, the water-borne solution proved itself hygienically and aesthetically in the setting in which it was invented. Its success over time was remarkable. What was also remarkable was that the sanitary reformers, in pushing for the spread of water-borne solutions, managed to make sewers, water closets, dung-heaps and 'excrementitious effluvia' part of the discourse of the Victorian age, even in daily newspapers and magazines read by polite society. The opening of Bazalgette's southern intercepting sewer outfall into the Thames east of London on 4 April 1865 was attended by the Prince of Wales, Prince Edward of Saxe-Weimar, the Lord Mayor of London, the Archbishop of Canterbury, the Archbishop of York and 500 other guests, who dined on salmon while the city's excreta gushed forth beneath them.[45] In the 21st century, celebrities such as Bills Clinton and Gates, not to mention a long list of singers, actors, royalty and religious grandees of similar renown, are happy to attach their names to campaigns on water, but rare are those to have identified themselves unreservedly with the need for sanitary advance.

Although the technology and the great reforming figures have taken pride of place in accounts of the 19th-century sanitary story, a vital thread of the long campaign consists of complex issues of finance and governance. These considerations are equally prominent in the debates surrounding environmental sanitation today. The effective operation, maintenance and sustainability of new facilities and infrastructure, without which all investments will be wasted, are more important than whether their initial costs of construction can be borne. These in turn will depend to a considerable extent on customer demand and appreciation – expressed in purchasing power – for new household improvements. Where expensive and technologically complex facilities are likely to fail these tests, it will not be surprising if initial investments are not forthcoming, whether these take the form of private capital, subsidized loans, or forms of aid or public provision.

In 19th-century Britain, the need to take more decisive and pervasive action on the sanitary front helped to spearhead radically changing attitudes about the role of government generally in contributing to the public good. Chadwick was a follower of Jeremy Bentham, a political philosopher who rethought the role of legislation and government in the modern world (he was, for example, the first person to come up with the idea of a ministry of health). In keeping with Bentham's ideas, Chadwick was convinced that the state must intervene forcefully to create the necessary public bodies, enable the sanitary works to be funded, lay down the regulations and standards, and enable the whole package to be enforced. He became intensely frustrated when this turned out to be so difficult to achieve. Eventually, though, such ideas were to prove their point. In the process, the whole system of British municipal administration was transformed, the theory and practice of public health established, and the business of water supply, street cleaning, drainage and excreta disposal removed from the control of private enterprise and taken into the public domain.

When the ground-breaking Public Health Act of 1848 was passed, it did not, however, contain the powers of enforcement that Chadwick had wanted. Towns were expected to appoint Inspectorates of Nuisances and Medical Officers of Health and invite the General Board of Health to send in their surveyors and devise a local sewerage plan. Take-up was slow, and where the board did have powers to enforce, it was reluctant to use them. There was widespread resistance to the interference from the centre that the sanitary mission represented, and this was echoed both in political parties and in small-minded municipal corporations up and down the country. Until 1867, when the electoral system was opened up, these were dominated by narrow-minded men of small property, wholly ill-equipped to respond to the modern demands of rapid urbanization. Those who paid the rates – the propertied classes – were unwilling to bear the expense of sewerage and the extension of services such as improved water supplies and solid waste disposal to the meaner parts of town, and the poor themselves were unwilling to bear the charges for connections that would be imposed. Only after 1870 did investment for municipal public health infrastructure really take off. Thus legislation in a number of areas and the opening up of the capital market to obtain loans on easy terms were needed before the necessary groundwork for the effective management of the various forms of waste – and other aspects of decent urban living – could be laid.

Even when laws and regulations were in place, enforcement was lacking. Cases brought against the perpetration of 'nuisances' were often lost. Doctors would be produced in court to say that stinking piles of dung were not a risk to health: until micro-organisms had been shown to convey disease, juries might be swayed by the lack of available proof. Corruption was common. In a case in Birmingham in the 1880s concerning polluted wells, the prosecution could not manage to persuade the court that they were injurious to health.[46] Lack of compensation hit the poor and discouraged cases from being brought. During outbreaks of typhus or cholera, bedding, furniture and household goods would be removed and destroyed. Dwellings might be fumigated with sulphurous acid gas for many hours, ruining wallpaper and floor coverings. Although legislative provision was made for compensation after 1881, it was rarely if ever applied. Magistrates would find that heaps of shit could not be removed, even when legally declared a 'nuisance', because they were the private property of the owner, and no provision in the relevant acts had been made for their seizure or for recovering the costs of carting them away.[47]

Before the shift could be made from 'dry' to water-borne sewerage, water supplies had to be provided in sufficient quantity. In most towns, private companies were set up to meet the demand, using for their sources rivers and underground wells regarded legally as in the public domain, but owning the pipes and charging for connections. As demand grew, most municipalities began to enter the water business. In this they were enormously helped by the availability of loans at favourable rates of interest from the Public Works Loan Commission. From the middle of the 19th century, they began to buy out the private companies and extend their networks of pipes and connections into the poorer parts of town.[48]

Gradually, reservoirs outside the large metropolitan centres were built, drawing water from considerable distances away. London, which was well served by artesian wells, was among the last authorities to take the ownership of water supplies into public hands, following repeated water shortages in the East End during the 1890s. By the end of the century, life in the cities had become as wholesome and healthy as life in the countryside. Even so, every commentator of the period remarks on the way the story progressed differently in different settings, depending on local systems of governance, topography, economic means, social and municipal attitudes, and the evangelizing spirit of city fathers. If this was the case in such a small and relatively homogeneous island state, how much more has it since been the case in countries within which are found far wider divergences in geography, settlement patterns, peoples, cultures, theories of disease, systems of local justice and administration, and economic means.

A long series of sanitary acts culminated in the Public Health Act of 1875. This was the work of the Chief Medical Officer to the government of the day,

Sir John Simon, another figure looming large in the sanitary revolution. This finally codified all the overlapping jurisdictions of Nuisance Inspectorates, Medical Officers and Boards of Guardians, and brought their various officers and staffs under the administration of Local Health Authorities. It laid the basis for public health so thoroughly that no change was required for a further 60 years.[49]

Finally came the impact for which everyone had waited. Mortality rates began seriously to drop. Between 1838 and 1854, the average age at death in England and Wales was 39.9; by the early 1880s, it had reached 41.9 and, by 1890, 44.[50] The different death rates between the classes were rarely assessed, but one urban medical officer of health calculated in the 1890s that the mean age at death for gentlemen was 60 and had been stationary since calculated by Chadwick in 1842, whereas that of shopkeepers and tradesmen had risen from 30 to 36 and that of artisans from 26 to 31.[51] The fall in general mortality between 1838–1847 and 1905–1914 was 37 per cent, with a high concentration of saving of life among children, young people and women in child-bearing years.[52] The advance of medical science, improved incomes, greater democratic participation, and a reduction in corruption and inefficiency in public life all played an important part. But the state- and municipality-driven sanitary revolution – in sewerage, street clearance, effluent treatment and plentiful water supplies – was the backbone.

Throughout the rest of the industrialized world, the process of urbanization during the 19th and early 20th centuries led to a similar public health revolution, in which underground drains and water-borne sewers were increasingly introduced. Many of the engineers and experts who had pioneered change in Britain became advisers to contemporary European city planners, Haussmann in Paris, for example, and to the colonial authorities, notably in India. Sanitarians in Europe were long divided into two schools of thought over the virtues of 'dry' and 'water carriage' methods of excreta removal, but eventually the chemists and hygienists put their weight behind the use of water as hygienically preferable. Systems of sanitation using dry conservancy and bucket collection were never tried in the US, where large-scale sewers were being widely introduced by the late 19th century. The last US city to banish cesspools was Baltimore, in 1879.[53]

When, in the 19th and early 20th centuries, the European powers began to develop systems of colonial administration in their Asian, African and American territories, the sanitary idea went with them. But the construction of sewerage systems was not on a par with the roads, railways, waterworks and other types of basic infrastructure colonial engineers were beginning to put in place. Efforts were made to introduce ideas of hygiene and sanitation to local populations, usually by means of a 'sanitary inspectorate' whose task was to visit towns and villages and make their inhabitants keep their streets and compounds clean. But construction of actual sewers or septic tank systems was confined to the urban 'cantonments' and 'colonies' inhabited by the administrative, military and professional classes whose membership consisted exclusively of 'their own'. Later on, when the colonial rulers and their civil officers withdrew, these were taken over by upper- and middle-class inheritors within the newly independent states.

With the exception of Mahatma Gandhi's protestation that 'sanitation is more important than independence', the need for the efficient and hygienic disposal of human excreta has not since become a matter of major public campaigning or moral reform in the world at large. When public health issues began to take on more importance in the development portfolio of the post-colonial world, the emphasis tended to be on medical technology for disease control, such as immunization. When sanitary issues began to enter the frame in the 1970s and 1980s, the concern was mostly for 'safe drinking water' as the route to better health. Only very recently have international experts and programmes begun to insist on separate and targeted action for changes in hygiene and sanitation. For far too long, the extraordinary accomplishments of the 19th-century generation of sanitary heroes had succeeded in putting excreta, its hazards, and its removal from homes and streets out of sight and out of mind. But today, finally, burgeoning urban populations, high levels of water and soil pollution, squalor in slums and crowded settlements, municipal mismanagement and need for reform, and epidemics of diarrhoeal disease posing serious risks to thousands of lives are pushing these issues back up the agenda.

Once again, solutions – technological, administrative, legislative, social and political – to a major worldwide sanitary crisis are needed. But the responses so far have been lacklustre and thin. Ironically, the country which pioneered the first sanitary revolution became, a century later, the brand-leader in turning its back on the lessons of those days. In 1989, the UK government privatized its water and sanitation industry, dismantling a public service that had lasted for six generations, together with the entire canon of legal and local controls representing the combined wisdom of governmental entities ever since Chadwick first articulated the gospel of public action to bring about sanitary reform. This privatization took place in the face of mass popular and political protest – protests that reoccur sporadically in the face of leakages, shortages, and stories of chief executive officers and share option holders pocketing millions in takeover deals.

2

Runaway Urbanization and the Rediscovery of Filth

Previous page: A boy collects plastic bags and bottles to sell, from debris floating in a low-lying slum area in Pasay City near Manila, capital of the Philippines. Some 6000 people live in shacks lining this stagnant pond.

Source: UNICEF/Giacomo Pirozzi

Before the sanitary revolution, cities and towns represented epicentres not only of power, wealth and civilization but also of dirt and pestilence. Faced with a life-threatening epidemic, townspeople who could afford to flocked to the countryside. In 1665, for example, the English diarist Samuel Pepys records as a matter of course the removal from London of King, court and gentry on the arrival of the plague from Amsterdam. Good air was country air, and people of means who 'ailed' or were actually sick went away to hills, dales, mountains, lakes or seaside to recover. The early tourist industry, with its mountain resort and coastal spa destinations, similarly responded to the quest among an expanding middle class for a healthy complexion and robust constitution.

The filth and disease with which town life was intimately associated was exacerbated by early industrialization – by the dirt and pollution generated by industrial processes as well as by the multiplication of populations living in cramped urban spaces. This remained the case until industrialization itself came to the rescue, building sewers and drains, paving the streets, developing a consumer market for perfumed soaps and bathroom ware, and banishing cesspools and 'public nuisances'. Although the squalor and potential for epidemic infection were augmented by the density of urban as compared to rural settlements, the greater numbers and proximity of town populations also had advantages in terms of it being easier to provide them with facilities for an improved quality of life. Their congregations of people, which included most of those who were powerful and wealthy, were easier to reach, and therefore easier to tax and also to provide with hospitals, drains and rubbish collection. Thus, over time, industrialization gradually reversed the conventional wisdom about healthy and unhealthy living environments. 'Town' became wholesome and salubrious. The naturally superior airs and fresher waters of the countryside were quickly forgotten, and the hygienic supremacy of urban life established.

In the early 21st century, the rural diaspora which originally transformed the industrialized world arrives at a Rubicon. The moment is expected sometime in 2008, when humanity will become a mainly urban instead of a mainly rural species. The last few decades have witnessed a headlong rush to urbanization all over the developing world. Cities in China, India, Brazil, South Africa and elsewhere outshine each other in architectural magnificence as their populations and economic outputs soar. But there is a dark side to this metropolitan transformation. The process has been even more chaotic and unplanned than its historical predecessor in Europe and North America. It took London over a century to multiply its population by seven, to 7.3 million in 1910; today, cities such as Dhaka, Kinshasa and Lagos have multiplied their populations by 40 since

1950.[1] In 1950, only New York and Tokyo had populations close to 10 million. Today there are 20 such 'megacities', most of them in Asia, Africa or Latin America. Over a million newborns and migrants are added every week, and it is predicted that by 2030 four in five urban residents will be living in the towns and cities of the developing world.[2]

The majority of these urban dwellers will not be living in apartment blocks with bathrooms and flush toilets, nor will they be using large amounts of power or otherwise causing mayhem with their carbon footprint. The most serious environmental aspect of this rapid urban growth is that most of these people will be poor, and their homes and living environments will be well below any acceptable standard that humanity should have to endure. Their living space – if they have living space and are not camping on pavements as do up to 65 million homeless in India alone[3] – will be crowded, full of debris and shit, much as were the rapidly growing towns of 19th-century Europe. So far, 'industrialization' has not come to their rescue, nor has it really even tried. Rather, the application of industrialization as the gauge of human progress has rendered these populations semi-invisible, even though their numbers are proliferating at a far faster rate than those of middle- and upper-income urbanites. Since the vast majority of people in developing countries who are wealthy also live in an urban and indus-trialized setting, average statistics make their urban populations appear misleadingly well off.

Rural people are often understood to be 'poor' because they lack access to 'improved' – essentially, manufactured or engineered, and therefore saleable and profitable – goods and services. But they usually have access to natural products husbanded from the environment. They also have space, and in matters of waste disposal and hygiene they have developed behaviours which, if not ideal, at least pose health risks that are relatively manageable. For example, many rural popula-tions use specially set aside areas outside the village or in the bush for sanitation purposes, and they have particular rules of cleanliness and hygiene, including over materials – herbs, sticks, leaves, pastes, water from this pond as opposed to that – to be used for toilet and toilette. 'Unimproved' though these systems may be, they have the advantage that they require no recourse to cash, unlike all 'improved' hygiene and sanitation methods. For those in the towns, natural systems are far less accessible, and since the residents of pavements, slums and shanty-towns have barely joined the cash economy, 'improved' amenities are not readily accessible either.

Unfortunately, the availability of 'improved' amenities in towns leads to the assumption that all those who live there benefit from them, which is clearly far from the case. Meanwhile, life in the 'unimproved' countryside is assumed by definition to be more deprived, less conducive to wellbeing and much more prone to infectious disease – an idea our forbears would have found extremely odd. Despite its over-simplicity, this idea of acute poverty and insanitariness predominating in the countryside as opposed to the town is somehow hard-wired into the development brain. The fact that a more urbanized country is also a more industrialized and richer country, and therefore that its life expectancy rates are higher and child mortality rates lower, is taken as proof that rural areas are where dirt, squalor and unhealthiness exclusively reside.

The reality is that a huge proportion of the world's seriously poor live in towns and cities in circumstances of extreme deprivation, in environments which far outstrip those of their rural counterparts in sanitary disadvantage and disgust. At least a billion people – one sixth of the world's population – live in crowded and dilapidated tenements or in shanty settlements spreading like tumbleweed on urban peripheries.[4] The numbers are vague because it is often unclear where the countryside stops and the town begins, or who is included and who – for example the homeless, the 'vagrant' and those illegally squatting on waste ground – left out. Of the 60 million people added to the world's towns and cities every year, most occupy the neighbourhoods known as townships, *bustis*, *barrios*, *bidonvilles*, *favelas* and, simply, slums. Some authoritative observers believe that the proliferation of urban poverty is set to become the most significant, and politically explosive, problem of the 21st century.[5]

The histories of these neighbourhoods vary, from economic downturns in inner cities to conflicts or droughts in the countryside that leave victims with nowhere else to go, collapses in agricultural prices, eviction from their lands by builders of dams and other infrastructure, outward encroachment onto farmland at the urban edge, or the absorption of villages into a new kind of urban space in which the town has migrated to the countryside rather than the other way around. The resulting settlements are classified as 'low-income areas', but they are very different in age, appearance, topography and size. They may constitute tiny enclaves of a few dozen families, or thousands of people camping out in a cemetery or public park. In low-lying areas of Asia, they may occupy derelict or swampy land, in which no-one wants to invest because it is hazardous, infertile or suffers from topographical blight. Such land may be extremely difficult to drain, and even in near-desert conditions, surface and groundwater contamination becomes a seasonal peril. Huddled colonies also occupy wasteland beside railway tracks and along roads, from which eventual eviction is guaranteed. In towns built on watercourses, they live on muddy river banks with the flimsiest shacks built out into the stream. In much of Central America, they perch on vertiginous hillsides liable to subside in storms, earthquakes and other natural disasters.

The average proportion of developing world city populations living in such neighbourhoods is thought to be between 30 and 40 per cent;[6] in some it rises to a half or even higher – 80 per cent in Dar-es-Salaam, for example.[7] Most lack infrastructure of the most basic kind, including a hard surface on streets or lanes, power lines, street lighting, covered drains, sewers and piped water. In places where corrugated iron, bamboo, torn-up packing cases and plastic sheeting are common building materials, there is no space, privacy or personal security. In most such shanties, five or six people typically share a single room.[8] Intimate behaviour, including washing, changing clothes and going to the toilet, is deeply problematic, as is laundry and getting rid of wastewater. In parts of the urban periphery of Dakar, Senegal, women charge US$0.50 to take one basinful away – which makes keeping clean and well turned-out a hugely expensive affair for people whose income is derisory (Figure 2.1).[9]

In Europe and North America, the municipal planners, social reformers, architects and engineers pursuing the industrial transformation of economic life in the 19th and early 20th centuries had time and resources to address the challenges posed by the huge influx of 'the labouring poor' into urban spaces. Political enfranchisement, civic enlightenment, workers' organizations, social reform, establishment of health clinics, and the expansion of consumer tastes

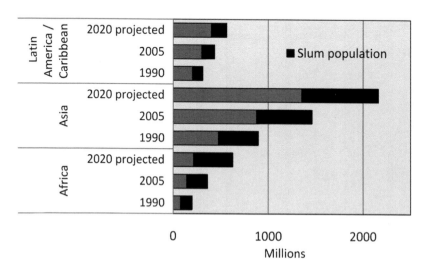

Figure 2.1 *Urban populations in developing regions*

Source: UN-HABITAT (2006) *The State of the World's Cities 2006/7*, Earthscan, London

and products were all part of the metropolitan makeover. The equivalent trans-formation in the developing world has been radically different. Infrastructural improvements stop at the threshold of the slums. Their inhabitants are not repre-sented at city hall, they are not employed in the economic mainstream, and they do not belong to trade unions or similar organizations able to fight their corner. Water, drainage and sewerage utilities are beyond their reach, and street lighting, rubbish collection and functioning stormwater drains are unknown to them. Moreover, they have no governance experience beyond the family unit, little education or skills, and no purchasing power with which to join the consumer world. Unlike in the village, there are few ties of family and social solidarity to offer a safety net when times are hard or the family stretched to breaking-point. On top of this, in town every necessity – from food, fuel and water to building materials and dirt removal – has to be paid for with cash.

Many towns and cities have gone out of their way to discourage new arrivals, treating incomers as illegals, transients or 'marginals' who have strayed temporar-ily into town and do not belong. Their perceived impermanence has been used as an excuse not to provide them with services, which are only laid on in newly planned or 'improved' areas. The argument that amenities in slum areas would attract more rural indigents has justified their neglect. Extreme measures – razing shanty-towns without warning, mass evictions to 'resettlement sites' – have been frequently used against them. In Dhaka, for example, between 1989 and 1999, 45 slums were demolished, some of them twice, by force and with police violence, leaving millions of families homeless.[10] In 2005, the city slums of Harare and other Zimbabwean towns were destroyed in an operation called Murambatsvina, meaning 'clear the filth'.[11] Thus people compelled to live in squalor are even personified as squalor. Efforts they make to improve their environment may be outlawed, as, for example, is the construction of any kind of pit toilet in urban Zimbabwe.[12] In many countries, when cholera or some other life-threatening diarrhoeal epidemic erupts, it is denied. It seems a short step to the banning of defecation itself.

In the 1970s, when the demographics of 'exploding cities' first drew the world's attention, various donor and local organizations began to take up the cause of improving slum communities by promoting a strategy of 'urban basic services'. This was a low-tech, low-cost alternative service-delivery, infrastructure-development and income-generating approach for low-income communities based on their participation. Communities organized, formed women's groups and water and sanitation committees, paved pathways, sank wells, and dug pit toilets with a little financial help and some technical advice. In some countries – for example Kenya, Zambia, the Philippines, Pakistan, Brazil, Haiti and Bangladesh – urban basic services became an important component of programmes of cooperation operated by UNICEF, among others, during the 1970s and 1980s.[13] But outside the handful of municipalities willing to consider the situation of citizens in far-from-leafy suburbs, neither the authorities nor other international partners took up these approaches with alacrity. Replication on any useful scale failed to materialize.[14] UNICEF itself later downgraded this area of its work, as did other international donors, although some urban basic services programmes continued during the 1990s, for example in Guatemala, the Philippines, pockets of African cities such as Kampala and Ouagadougou, and Bangladesh.[15]

Today, a number of international and local NGOs are indirectly concerned with urban issues in the context of micro-credit and social support for women, child labour, children living on the streets, and child sexual exploitation. The extreme level of violence among young people in urban Brazil and Central America has attracted much attention. But concerted efforts to deal with the slum conditions that generate violence and deplete family cohesion have been lacking. With certain notable exceptions, for example in urban Philippines,[16] usually where slum populations have managed to organize themselves and represent a political force, governments and municipalities rarely ask for assistance for their urban poor, and the donors of today – the NGO community excepted – rarely offer it. From 1970 to 2000, the amount of urban development assistance from the major donors was estimated at US$60 billion, just 4 per cent of total flows.[17]

Housing, water, sanitation and other services to improve conditions for the urban poor received only 11 per cent of total lending from the World Bank between 1981 and 1998; they are also neglected in the Bank's 'poverty reduction strategy papers', today regarded by the big donors and their national governmental partners as key planning instruments for social improvement. This is partly because for some years there was a misguided illusion in the international community that private sector participation would step in and substitute for public investment in housing, and water and waste infrastructure. Unfortunately, as far as the very poor were concerned, it did not. Meanwhile towns and cities large and small are where the poorest and most alienated people are increasingly to be found. If their right to a share of services is not respected, the crisis of squalor, unhealthiness, violence and social resentment will only get worse. In effect, a similar crisis of social disorder and political unrest as that which inspired sanitary reformers such as Edwin Chadwick in 19th-century Britain is being played out on the streets of many rapidly growing and grossly deprived urban environments of today. Perhaps it will take the making of an equivalent causal link in the policymaking mindset to force environmental sanitation for urban areas higher up the public agenda.

The very fact of people's determination to vote with their feet in favour of urban life, and the resilience they display in the face of municipal neglect and lack of jobs, security and tenancy rights, shows that slum people are resourceful. Many are motivated to improve their situation and the prospects in life for their children.[18] Just as in the earlier story of urbanization, a critical area of people's need is a sense of control over the mud and mess of alleyway life. They need a safe and hygienic method of excreta disposal, plus places to wash in privacy, and water for keeping themselves, their homes and their children clean. A new kind of sanitary revolution in *favelas* and *bustis* might provide the same impetus to improved urban living as its historical precursor. But in order to make that happen, a much greater degree of public recognition is needed about the lamentable state of sanitation in urban back streets and shanty-towns, and the fact that this is not a temporary phenomenon but a permanent feature of social transformation requiring suitable responses. Also needed is a greater appreciation of how desperately many urban inhabitants need and want change, and that 'neighbourhoods fit to live in' might go far to reduce high rates of crime and social tension.

The modern story of international sanitary reform in the post-colonial era begins with the UN conference on water in 1977 at Mar del Plata, Argentina, and the declaration of 1981–1990 as the International Drinking Water Supply and Sanitation Decade (IDWSSD). The impulse behind those who pushed hard for this to happen was the knowledge that urban populations were rapidly expanding and that conventional responses to sanitation – flushing toilets, household connections, sewers, treatment plants and the rest of the industrialized world's public health infrastructure – were impracticable and unaffordable for most inhabitants. In the 1970s, 'exploding cities' was a *cause célèbre*, with radical voices such as architects John Turner and Jorge E. Hardoy arguing for alternatives in urban development. As defenders of the urban poor, they presented a case for not tearing down slum and squatter housing, but recognizing it as a vigorous and creative form of self-help. As a result, organizations such as the World Bank downgraded their 'sites and services' type interventions and abandoned public housing and subsidies for urban infrastructure as ideologically outmoded.[19] Inadvertently, the resourcefulness, courage and capacities of slum dwellers were co-opted as a pretext for the withdrawal of public resources by governments and donors.

At this time, the basic framework of anti-poverty thinking was switching away from Dickensian visions of slums and towards the transformation and commercialization of rural life – partly as a way to stop urban migration in its tracks. The Water Supply and Sanitation Decade coincided with this trend, and within public health engineering the predicament of the rural 'unserved' began to take precedence over that of the urban. The much larger numbers of those in rural areas of the developing world who had no water or sanitation facilities at all grabbed almost all the Decade's attention. In 1988, when 'coverage' was reviewed, sanitary progress in urban areas had encouragingly reached well over 60 per cent, while in rural areas it was static or getting worse at less than 15 per cent.[20] The major achievement of the Decade was in low-cost rural water supply, and when it ended, the superior claim of rural people for water and sanitation had somehow become entrenched.

The Decade established as the benchmark of progress a numbers game around 'coverage' figures for services, setting up an engineering and construction race to install ever larger numbers of taps, drop-holes and flushes in different environments. Ever since, the spotlight has continued to shine more brightly on rural service shortage than on urban. Today's official statistics on sanitation coverage – and water supply, for that matter – suggest that only a minority of urban dwellers are 'unserved'. Of the 2.6 billion people described globally as 'without improved sanitation', the vast majority – 2 billion – are in the rural areas, with a much smaller number – 611 million – in the towns (Figure 2.2).[21] This figure is reckoned to be an underestimate by today's leading authorities on slum and squatter settlements, who believe that the invisibility of the poorest and most deprived urban populations in data collection obscures the fact that residents in the worst living environments have nothing resembling adequate sanitation (Table 2.1).[22] As they are living in settlements which officially do not exist, such people frequently do not get counted in surveys or censuses. But whether the figure is closer to 600 million or 900 million, as urban specialists suggest, it still implies that efforts must be overwhelmingly targeted to rural areas if the Millennium Development Goal (MDG) for sanitation is to be reached. The tone of MDG reporting – that globally there are four times more unserved rural dwellers than unserved urban dwellers – unintentionally tends to reinforce this impression.

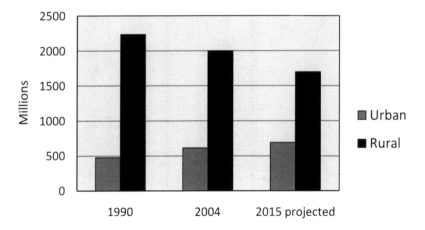

Figure 2.2 *Global population without access to improved sanitation*

Source: WHO and UNICEF (2006) *Meeting the MDG Drinking Water and Sanitation Targets: The Urban and Rural Challenge of the Decade*, WHO/UNICEF Joint Monitoring Programme, Geneva and New York

However, any perspective that downplays the problems of sanitation in low-income urban areas is unrealistic. It is unrealistic for various reasons, but one of the most important is that sanitary problems – both the difficulty about where to 'go' and what to do with the noisome result in places with no organized systems for excreta removal – are much more acute in crowded urban spaces than in rural spaces. It is also unrealistic because the reported figures for sanitation do not properly reflect what is happening in squalid and irregular 'colonies' where people are living without official sanction or notice.

According to country reports to the international monitoring programme jointly run by WHO and UNICEF, 53 per cent of African urban populations had 'improved' sanitation in 2004.[23] This figure was definitely a more accurate assessment than the 84 per cent claimed by their governments in 2000. But individual case studies suggest that these synthesized figures still do not capture the typical experience of many urban dwellers (Figure 2.3). Some municipalities

Table 2.1 *Urban dwellers lacking adequate provision of sanitation, by region, 2000*

Region	Number
Africa	150–180 million (50–60%)
Asia	600–800 million (45–60%)
Latin America and the Caribbean	100–150 million (25–40%)

Source: David Satterthwaite and Gordon McGranahan (2007) *State of the World 2007: Our Urban Future*, The Worldwatch Institute, Washington, DC, and W.W. Norton, New York

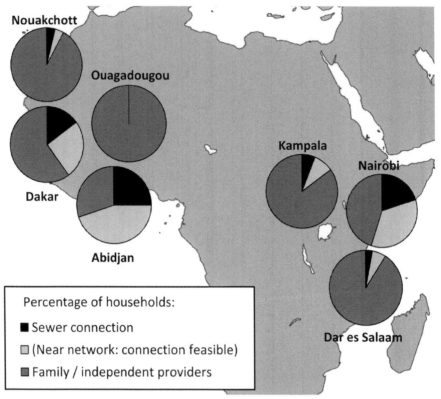

Figure 2.3 *Access to sanitation services in seven African cities*

Source: Bernard Collignon and Marc Vézina (2000) *Independent Water and Sanitation Providers in African Cities: Full Report of a Ten-Country Study,* Water and Sanitation Program with other partners, The World Bank, Washington, DC

assert that 100 per cent of their citizens are graced with flush toilets. This is the case in Mbare, a densely populated inner-city area of Harare, Zimbabwe. But a recent study found that, in one Mbare neighbourhood, up to 1300 residents shared one communal toilet with six squatting holes, an unreliable flush and no electric light for use during the night.[24] The most cursory visit to the slums of many African towns and cities shows that any claim of high or near-universal coverage has omitted to count most of the populations living in squatter or ramshackle townships, even some that have been there for decades, whose residents are those who most desperately need and want somewhere decent to perform their bodily functions. And overstatements are not confined to Africa. Many countries elsewhere – Nicaragua for example – report sanitation coverage figures in the region of 85 per cent and above, which no competent observer believes.[25] To muddy the water still further, the term 'sanitation coverage' has in the past often been confused with 'toilet usage'. Although household surveys now try to capture the difference, facilities which fall into disrepair within a year or two of construction may be inadvertently included in 'coverage' because they were being used when the survey was undertaken, even if they could not stand the test of time.

When it comes to water supply, the problem of 'inadequate' service is at least publicly and widely acknowledged. Stories about women queuing for hours at the standpipe, or walking long distances with a heavy container on their heads, are part of the narrative about the predicament of poor women concerning household water. One reason the toilet predicament is less well known – even to the extent that it is falsely assumed that demand for urban sanitation does not exist – is because there is no equivalent narrative: the subject is taboo, the story never told, the question rarely put. Some researchers have begun to overcome their squeamishness, and information is now beginning to trickle out. These researchers are usually women – illustrated by a recent study by Deepa Joshi – as are those they speak to, an acknowledgement that women tend to feel the absence of sanitary facilities more acutely than men. This is especially the case in societies where women do not move around freely in public, as in South Asia, where shame and humility surrounding lack of sanitation facilities can be acute.[26]

An account from Pune, India, comprising many interviews with women living in the slums collected by an NGO called the Society for Promotion of Area Resource Centres (SPARC) illustrates the point:

> *We used to go to the toilet near the river side. The insects used to climb up our legs. Or I went in the bungalow where I worked, or we went to defecate under the bushes. Then in the elections, Qazi Saheb [a local politician] came and arranged for taps. After this each house had a tap, but there was no provision for toilets. Even today the toilets are as they are. It takes one to one and a half hours to use the toilet. And even now, insects climb our legs.*[27]

Women in parts of Mumbai who live near the railway tracks use them as a toilet between midnight and 4.00am. 'We sat between railway tracks, and if a train came, we used to jump onto the other track. There were frequent accidents and every week or so, someone used to get killed.'[28] Between 20 and 50 per cent of people in Indian towns and cities live in slums and shanty-towns, and many of them suffer a similar predicament. According to the best available WHO and UNICEF statistics, between 14 and 26 per cent of India's urban inhabitants have no toilet of any kind.[29] Small-scale studies in urban areas of Asia and Africa throw up countless examples where large proportions of the urban population have no facility at all, or are obliged to rely on buckets or foul latrines used by scores of others.[30] Children may freely use the open air, squatting over drains or gutters, not even attempting to find some kind of privacy. Children's faeces are particularly pathogenic, and this behaviour may help to explain why infant and child mortality rates remain high in many urban areas, even where some form of sanitation has been provided.

Although there has recently been a much more systematic effort to gain an accurate picture of sanitation in the poorer parts of towns, there are real problems to overcome. One is the reluctance of officials to acknowledge the presence of city dwellers who are unregistered, 'vagrant', occupying illegal land or living in chaotic alleyway warrens. Another area of contention is whether the results of household surveys, on which the compilation of international statistics mainly relies, can really provide an accurate picture of facilities describable as accessible, adequate and safe.[31] Yet another concerns exactly what constitutes 'improved' sanitation, and by whose estimation and criteria it represents a sanitary advance on the unimproved alternative. In a great number of minds, including those of most laypeople and many officials sensitive to the municipal or national reputation, the only 'improved facility' remains a flushing toilet. During the 1980s, this was the only definition of 'adequate' sanitation recognized in Brazil, for example, and many accounts of sanitation are still written as if nothing else qualifies.

In fact, if everything bar flushing toilets was eliminated from the definition of 'improved', the total numbers of those in the predicament of lacking sanitation would rise to over 4 billion.[32] Among these, at least 1 billion would be in towns or in quasi-urban settlements currently classified as rural. The WHO/UNICEF monitoring programme has recently tried to sort out the muddle of what is 'adequate' or 'improved' or constitutes 'coverage', and in its most recent report sounded a warning that more attention is needed for urban areas given their rapid population expansion and proliferating slums. The latest report predicts that, on current trends, the number of urban dwellers without sanitation will increase significantly from the baseline year (1990) by 2015, while the equivalent number of rural dwellers will decrease by 25 per cent (Figure 2.4).[33] The monitoring programme is charged with measuring progress towards the Millennium Development Water and Sanitation Goals, and, with the best will in the world, the task is highly problematic. However, since MDG attainment is currently driving policy in the global sanitary world, all definitions and their nuances matter.

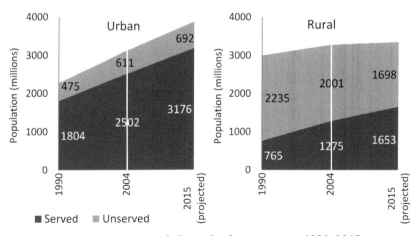

Figure 2.4 *Trends in sanitation coverage 1990–2015*

The urban figures shown here as unserved are lower than other estimates from non-JMP sources. They do not include urban populations in shanty-towns and unplanned settlements currently uncounted because they are 'illegal'.

Source: WHO and UNICEF (2006) *Meeting the MDG Drinking Water and Sanitation Targets: The Urban and Rural Challenge of the Decade*, WHO/UNICEF Joint Monitoring Programme, Geneva and New York

Where there is no toilet facility in the house, and no decent public facility or place set aside in a sheltered corner of the urban neighbourhood for the purpose, people – or anyway women and children – usually 'go' in their own homes. Some keep the outcome in a bucket, later to be poured into an open drain, where it finds its polluting way into a waterway. Alternatively, they may defecate onto an old bread wrapper or into a plastic bag, and 'wrap and throw' the little bundle. It flies onto the nearest available dump or into a communal skip (hence the term 'flying toilets'), or ends up on a derelict plot, overtaken by communal rubbish, where dogs and pigs go to scavenge. In the shanty-towns of Nairobi, Addis Ababa or Lagos, or indeed any number of other African cities, and in parts of Asia, 'wrap and throw' is the unimproved form of sanitation used by millions of people. For many, it is preferable to sharing a filthy and overflowing public toilet with scores of other families. The latter is unsuitable for small children at any time, and, where security is lax, alcohol consumption high and violence common, it is unsuitable at night for girls and women.

Just as municipal authorities may describe their population as having access to a water supply when there is only one unreliable tap stand every few hundred metres, they may describe them as having access to sanitation because some kind of toilet installation has been erected in a certain neighbourhood or in a number of households within the recent past. 'Improved' often simply means 'constructed'. In the WHO/UNICEF definition, it means that the toilet confines excreta effectively and is thus more-or-less microbiologically safe. But in the case of excreta, the term 'improved' is not really very meaningful to the consumer unless the toilet is congenial and able to eliminate nasty sights and smells on a lasting basis. This is where the motivation of public health officials for developing facilities, and the motivation of household members for using them, may not always coincide. Service back-up may also be lacking. If a toilet overflows in the rainy season and pours its effluvia all over the modest living space, it cannot be described from the point of view of the user as a permanent 'improvement' on visits to a secluded spot or even a bucket whose contents can be thrown away.

In every known sanitation programme where toilets have been built without first establishing whether people wanted them or what type they would prefer, some if not most are never used. In crowded urban areas, public toilets whose management and maintenance has not been given the same attention as their initial construction may quickly become disgusting and abandoned by customers. Coverage figures for 'improved' sanitation therefore not only sometimes give a distorted picture of the ongoing sanitary predicament of poorer urban families, but may have little relationship with what, actually, people do to meet their personal waste expulsion needs, or with what they would choose to do if a choice was available.

The optimal sanitary device is undoubtedly a flushing toilet whose water supply is piped in and whose waste is discharged into a sewer or septic tank. Its public health advantages are reinforced when water is also directed to a tap and basin for hand-washing after use. But its most desirable attribute is that it reduces unpleasantness. People think of their WC not as a health aid, but as a device for dealing with necessary bodily functions. Other considerations rate much higher with them than reduction of disease.[34]

A flush toilet has advantages over alternatives. It takes up little space – not much more than a bucket: a major plus in a congested urban setting. And where the flushed waste is totally removed from the scene without householder effort, this is a luxurious extra. If the output is later cleansed by passing through a treatment plant, ensuring that pathogens are kept separate from rivers and underground streams where they might contaminate the water supply and pollute the environment, the public health advantages are supreme. Thus convenience, congeniality, and the promotion of public and environmental health converge in the well-sewered flush toilet.

For the majority of people in the urban developing world, however, this kind of toilet and the associated public health engineering infrastructure of the conventional kind is unobtainable, for all the financial, technical, topographical and political reasons already explored. For some of them, the pared down, cheaper versions of water-borne sewerage that have been developed and success-fully installed in urban neighbourhoods in Pakistan, Brazil, Bolivia and other parts of Latin America show much potential for low- to middle-income commu-nities in towns and cities with suitable topographies and the necessary political commitment. However, although these could be applied far more widely, they are not going to provide the solution for most slum and shanty-town popula-tions. Even in better-off parts of Asia and in most of Latin America, they are only appropriate in a *barrio*, *favela*, or *busti* that is accepted as a permanent fixture on the urban landscape, and where wooden crates or bamboo as housing materi-als have already given way to concrete block and brick, water is laid on, and the general air is on the up – what are known in Tegucigalpa, Honduras, as 'devel-oping' communities (*en desarrollo*). In the informal shanty-town of packing-case, cardboard, flattened tin and other fragile materials, or in areas with swampy or unstable soils, simplified sewerage is not going to work.

So expanded sanitation coverage – defined in technological terms – will continue to mean different types of sanitary receptacle and waste removal systems in different town and city environments. In many urban slums, it will have to include 'on-site' solutions, a euphemism for pits and septic tanks. In areas where there are space limitations or where people are living in shacks or on the streets, it will have to mean public toilet blocks, also with pits or septic tanks if sewers are nowhere nearby. These different types of toilet are far more numerous in their small and large refinements than most people would begin to imagine (see Chapter 4). They can be 'improved' or 'unimproved' depending on their construction materials, pit depths and linings, emptying procedures, whether they are 'wet' or 'dry', and all sorts of other variables such as whether small children might fall through the hole. Their costs vary accordingly, ranging from US$12 (lowest per head price for a communal facility serving 50 people) to upwards of US$350 for ecological (composting) varieties (Table 2.2).[35] Some types of on-site sanitation are undoubtedly better suited to the countryside, with its larger spaces and fewer tenancy problems, and in congested town quarters the potential of pits for polluting the groundwater – as happened in 19th-century London – needs to be given extra consideration.[36] 'On-site' sanitation does present risks for urban areas, but these risks have to be tackled because of the lack of any possible alternative in many neighbourhoods.

Table 2.2 *Different sanitation options and their unit costs*

Type of provision for sanitation	US$ cost per household	Comments
A water closet connected to a sewer or septic tank within each home plus piped water to the home for personal hygiene	400–1500	Unit costs rise a lot if provision is made for sewage treatment using conventional treatment plants and with high levels of treatment
Condominial sewers (e.g. Orangi Pilot Project)	40–300	With high densities and strong community organization and input, unit costs per household can compete with pit toilets
An 'improved' (eg VIP) or pour-flush toilet within each home	40–260	No need for sewers. These control smells better than unimproved pit latrines and limit or prevent insect access to excreta
Ecological toilets	90–350	No need for sewers. Provision for urine diversion with economic advantages for nutrient recycling but can add significantly to unit costs.
A basic latrine	10–50	No need for sewers. If well managed can be as healthy as more expensive options; unit costs may be lower than US$10 in some rural contexts.
Access to a public or communal toilet/latrine (assuming 50 persons to each toilet seat)	12–40	Effectiveness depends on how close it is to users, how safe to use at night, how well maintained and how affordable by poorest groups
Possibility of open defecation or defecation into waste material ('wrap and throw')	None	Obvious problems both for those who defecate and for others

Source: David Satterthwaite and Gordon McGranahan (2006) *Overview of the Global Sanitation Problem,* Occasional Paper 2006/12, UN Human Development Report, United Nations, New York

Different types of waste disposal system can be seen as belonging to rungs on a 'sanitation ladder', starting with whatever is regarded as the minimal improvement on going for a walk up the alleyway, over the footbridge, or alongside the shore or river bank. In some places this may be seen as 'wrap and throw' (an urban equivalent to covering excreta with soil, known as the 'cat method'), or a covered bucket; the next rung up might be a fabricated toilet 'slab' situated above a simple dug pit. But as incomes rise, as is more likely to happen in the town than in the countryside, tastes and consumer aspirations change. As in 19th-century Britain, position on the sanitation ladder is an indication of social attainment. The passage of upwardly mobile urban-dwellers towards a better livelihood and higher social status is illustrated by the presence of certain amenities – television set, refrigerator, cabinet, upholstered sofa and chairs. For many of those rising off the urban floor today, a high priority will be a decent toilet.

For the urban setting has one very important asset compared to the rural setting when it comes to sanitation. The nature of crowded urban living means that people do want decent amenities. No-one who aspires to be anyone in town chooses to live without a proper place to shit if they can afford one. The powerful assumption that people who are unfamiliar with toilets do not want them and will not use them may be applicable in some of the open spaces of the countryside, but this is much rarer in town. Here, resistance to the use of a toilet is more likely to be associated with its insalubrious condition, the number of other users and the state they leave it in, or, where children are concerned, the fear of falling in or some other nasty experience. Stories of slum dwellers spending large amounts of their shockingly low incomes on purchasing cans of water are familiar in cost analyses of urban water supplies and have been frequently used to demonstrate 'willingness to pay'. What is less often noticed is that urban people are also willing to spend money on having a toilet in their home. In a pilot scheme in three communities in El Alto, La Paz, supported by the United Nations Development Programme (UNDP)/World Bank Water and Sanitation Program (WSP), 97 per cent, 98 per cent and 92 per cent of the inhabitants, when asked after a sensitization campaign, wanted to connect their homes to a simplified sewer.[37] This level of demand is by no means exceptional.

There is also evidence that those who have no toilet in their home and no possibility of installing one are willing to pay for a 'convenience', especially for the main visit of the day. Accounts of the daily lives of pavement dwellers in Hyderabad show that they rate highly the use of decent toilets, bathing and laundry facilities even when they earn only around 50 rupees (US$1) a day. They willingly pay the substantial sums of two rupees (US$0.04) a day for the toilet and five rupees every few days for a bath, so as to defecate in private and keep themselves clean; mothers will even spend 20 rupees once a week to bathe all their children.[38] Most of the facilities they use are run by Sulabh International, a unique private organization in the Indian sub-continent whose founder, Bindeshwar Pathak, was a disciple of Gandhi's sanitary crusades. Pathak, who opened his first Sulabh toilet complex in Bihar in 1970, refused to accept the popular wisdom that uneducated Indians invariably prefer to defecate in the open air. He insisted that poor people in towns do want to use a proper toilet and are prepared to pay for it, and the toilet empire he has developed proves his point. But he also understood that the facility has to be clean, well managed and properly staffed and maintained. More than 10 million customers use his complexes daily in 1080 towns all over India, and 35,000 people are employed to manage them.[39]

Similar demand has been demonstrated in another Indian city – Pune. There are over 500 slums in the city, containing two-fifths of the 2.8 million inhabitants. In 1999, a visionary municipal commissioner, Ratnakar Gaikwad, decided to boost the provision of public toilets by inviting NGOs to make bids for contracts to construct and manage them. One of the NGOs to receive contracts was SPARC, an organization with links to two people's organizations, the National Slum Dwellers' Federation and Mahila Milan, a network of savings and credit groups for women pavement dwellers. The three organizations entered into a deal with the Pune municipality to build 114 toilet blocks, with a total of 2000 adult places and 500 smaller ones for children. Certain local contracts were handled by women community leaders, and the design of the blocks – with a living space for the caretaker and his or her family, good ventilation and special children's facilities – responded to community desires. Some included a community hall, encouraging the caretaker and customers to keep the area clean. The whole project was carried out with frequent interaction between the community and authorities; there was accounting transparency and full consultation on user fees.[40] The Pune experience – and similar initiatives in Nairobi, Dhaka and elsewhere – shows that plenty of demand exists, that it is more a question of the effort being made to tap it.

In the end, the question of what constitutes 'adequate' or 'improved' toilets, whether in a slum, a tower block, a public facility or a palace, has to be decided by those who use them, or choose not to use them, and who show the lengths they are prepared to go to in terms of labour and community work, as well as what they are prepared individually to pay for use in the case of a public toilet. Rejection of a particular facility or refusal to install a particular model in their home does not necessarily mean that all sanitation options will be unacceptable to potential customers.

In Britain, toilet usage took off in the 19th century because the water closet became a desired consumer item, and a market and promotional activity developed around sanitary ware. Rapid installation of flush toilets led to the pollution of rivers, and thence to major public investments in water supply, sewerage and treatment plants. The adoption of toilets was led by popular demand, even though the leadership and clout of the sanitation movement derived from the dynamics of public health and municipal reform. Unfortunately in the process, sanitary leaders so elevated the role of public health engineering and its capacity to remove shit, treat effluent and make dirty water clean enough to drink, that they took the issue of sanitary choice out of the province of individual action and into a 'we will fix it' public realm. They forgot about 'demand', and they forgot that people's preferences and behaviour play an all-important role in the take-up of any new device or service. In the mindsets of many sanitary engineers and municipal authorities today, technology and hardware are still at the centre of the frame, and they have merged the consumer item – the toilet – with the system of excreta disposal – the sewer or on-site storage unit – as if they were the same thing. Their aim is to construct as many toilet-cum-waste-disposal facilities in as many places as budgetarily possible. But however important the public health requirement to introduce sanitary systems of human waste confinement or removal, that is only one aspect of the issue.

The real question is not about how many installations of improved toilets can be constructed in the drive for universal 'coverage' and achieving the MDG, but about how to cultivate demand for hygienic living and all that it comprises. This includes toilets, of course, but also washbasins, reliable supplies of water, garbage disposal, and drains or soakpits to remove wastewater. It also requires a manufacturing industry that provides the necessary consumer items at costs people can afford, in shops or markets which they can reach, and in ways they can manage in terms of maintenance and repair. The other part of the challenge is how to flush out of a reluctant municipal machine the necessary leadership, resources and commitment to fulfil their public services role. And this means recognizing that, even when the excreta is going to be managed not by sewerage but by confinement in special pits, tanks or chambers, effective and affordable ongoing services will be needed. Urban pioneers such as Sulabh and SPARC have proved that there is a demand for decent facilities in crowded slum neighbourhoods if the kind of services provided match what people want. They have also proved that some kind of authority, private or public, has to shoulder responsibility for excreta management beyond the point of deposit.

Despite the scandalous lack of interest and investment in sanitation in poor urban neighbourhoods, a few approaches – most by NGOs or philanthropic entrepreneurs, but some in the public domain – have achieved great things. They show that with official responsiveness and community involvement, much can be done. Some might even be described as an investment opportunity, if they were so presented …

At the top of the 'sanitation ladder', below conventional sewerage, are the pared down, simpler, lower-cost sewer systems which were first built in the early 1980s. The best-known example is the Orangi Pilot Project (OPP) in Karachi. Here the leadership of architect and planner Arif Hasan, coupled with strong local community organization, managed over time to conquer an acute sanitary problem in an extensive network of informal settlements or slums (*katchi abadis*) enduringly and sustainably. The strategy was essentially one of mobilizing the inhabitants in each of the lanes in the *katchi abadis* to build their own small-scale lane sewer, all of which emptied – directly or via intermediary collectors – into a larger drain. So successful was this that before long the Karachi authorities agreed to build trunk sewers to replace the receiving drains in some of the settlements and connect the Orangi settlements into the main sewerage system. Local 'lane committees' become partners of the municipal authorities, cutting both administrative and construction costs.[41]

Since Arif Hasan first set up his NGO and began to motivate Orangi's inhabitants, nearly 96,000 households have installed toilets and connecting pipes and laid lane sewers with their neighbours, contributing their own resources to build and manage the systems. Altogether, the residents of Orangi have invested around US$1.5 million from their own pockets – a seventh of what an equivalent system would have cost the public health engineering authorities. For their part, the authority has provided major infrastructure and been willing to incorporate a different and decentralized sewerage system in its network – a not inconsiderable bureaucratic breakthrough. Orangi's achievement over 25 years has been extraordinary, a real success story in slum improvement, and proof that toilets and sewers can be the vanguard for all kinds of upgrading activity leading to better health and quality of life. Since the mid-1990s, its model has been successfully replicated in several other smaller locations in urban Pakistan, in and outside Karachi. This has come about with help from OPP and from the many enthusiastic donors who – because of its superlative community mobilization credentials – have beaten a path to OPP's door. Achieving 'replication' or 'scaling up' is the donor's nirvana, and OPP has managed it.

But for all its success, OPP's experience of 'replication' has not been entirely smooth. This is because of the need both for motivation and expression of demand and for community input and organization. In each case where the OPP replication has been successful – in other towns such as Faisalabad, as well as in other areas of Karachi – a local NGO has been trained for, and charged with, the motivational and community organization work, and the area of operation has been relatively small and manageable. Thus the only problem with the model is that engineers and bureaucrats cannot copy it unaided: they are not normally equipped to inspire or build demand. Actually, they are not usually interested in trying since – unlike in the case of manufacturers of bathroom ware and other household improvements – marketing ideas form no part of the training of engineers and administrators, who are taught to see themselves as 'masters of the best solution'. Large-scale replication of OPP has also proved difficult. The kind of ongoing inspirational leadership required to bring about community group action on the scale of Orangi, with everyone cooperating in a well-managed and well-structured democratic framework, is very difficult to replicate – in sanitation or any other context. It depends on extraordinary and outstanding personal qualities – sanitary heroes on a scale equivalent to Chadwick and Bazalgette – whose commitment covers decades.

The other part of the world in which low-cost sewerage has become renowned is Latin America. The system known as 'condominial sewerage' was first pioneered in Brazil in the 1980s. The 'condominial' tag is there because, instead of laying a sewer pipe in the road and connecting it to every house on an individual basis, the pipe runs from one house or dwelling to another as if they were in an apartment block. This substantially cuts the cost – by up to 70 per cent compared to conventional systems. In the early 1990s, the Water and Sewerage Company of Brasilia adopted this model because there was a massive sanitation problem in the city. Around 1.7 million people were without sewer connections, and conventional systems were unaffordable. After some experimentation, the condominial system was adopted for the whole municipality, poorer and more affluent. Between 1993 and 2001, an estimated 188,000 condominial connections were installed, benefiting 680,000 people.[42] Community involvement was also there, if not quite to the degree of Orangi. Households were invited either to dig the channels and lay the pipes themselves – under supervision – or to pay the utility to do the work for them. Households willing to install pipes in their yards and take on responsibility for maintenance had their fees reduced. Here, as at Orangi, the municipal authority was similarly persuaded to be flexible by embracing a decentralized network within its plan.

Subsequently, lower-cost adaptations of conventional sewerage systems have spread to many Latin American locations outside Brazil. But there is a tendency to deduce from these successes that the golden key is provided by technological innovation – smaller pipes, cheaper materials, cleverer underground routes. Plus, naturally, the cost advantages of people's contribution in labour or cash towards the installation process. However, while the importance of cheaper technologies should not be underestimated, the celebratory accounts of condominial breakthroughs have a familiar tendency to highlight the role of engineering and omit the importance of people's preferences, behaviour and demand. In a condominial system, a blockage requires a cooperative effort to resolve it: all householders along the line are complicit in the functioning of a small-diameter pipe, and if one household emits material which will not 'go down', there is a problem. In every community or 'lane', there are always difficult families. Lower-cost sewerage systems require cooperation in a way that conventional, centrally managed, systems do not. In a conventional system, each householder can call in the plumber for household blockages or, for external flow problems, summon the utility engineers. But with the smaller type of community-installed sewer, this is not viable. Community commitment and behaviour are decisive in whether such systems are really able to be delivered and maintained.

The story of low-cost sanitation in Tegucigalpa, Honduras, is instructive. More than a third of the city's population lives in makeshift communities – *barrios en desarrollo* – clinging to the steep hillsides surrounding the centre and its better-off suburbs. Their vulnerability is typical of low-income populations in towns and cities elsewhere in this disaster-prone region. In 1998 the torrential rains of Hurricane Mitch deposited all the buildings and people from some of the hillsides in huge mudslides at their feet. Residents of similar terrain today describe how they still tremble at night during heavy rains. But these rural incomers, who have tripled the city's population to more than 1 million in 25 years, and some of whom are now in their second or third generation there, had no alternative place to go to. Since they first began to set up shanties on slopes where horses grazed, they have gradually replaced their packing-case homes with solid brick structures. Mud paths have been paved and services gradually brought in. Water used to be carried by the women up perilous and slippery paths, but is now systematized by piped services or taken to regular collection points by tankers. Waste disposal consisted of throwing rubbish down the hill or into whatever stream passed by on its way to the filthy river at the bottom. As for toilets and sanitation, there was nothing.

In 1990, a special unit in the municipality, the Executive Unit for Barrios en Desarollo (UEBD), was set up to work with the National Autonomous Water and Sewerage Authority (SANAA) to help provide water facilities in these communities.[43] The UEBD's technical and financial requirements were initially underwritten by UNICEF, which provided support for its policy and approach. Later, in 1993, a foundation called Agua para Todos (Water for All) was established, with capital from the Chamber of Commerce and external donors, to provide revolving loans for the projects. Since 1990, 300 communities with over 200,000 inhabitants have been assisted.

From the outset, the programme was based on community mobilization and contribution. When a community requested a project, the first activity was for the UEBD social promoters to hold local assemblies, explain the process and financing, and prompt the election of a water and sanitation committee. The community had to accept responsibility for 60 per cent of the cost, for which a loan would be provided. And it also had to establish an action plan to work out how every household in the community would contribute to construction and maintenance and to meet a certain standard of sanitary behaviour – for example building pit toilets and not allowing standing water to collect and provide a breeding ground for mosquitoes. Persuasion to join in, and collection of dues, has to be done by the community; these are local governance issues and cannot be decided from outside.

By 1996, it became clear that 'improved' sanitation was also needed. As settlements became more crowded and plots were fully built over, pit toilets ceased to be a good solution – once full, there was nowhere to build a replacement. A low-cost simplified sewerage system was proposed, with small pipes and no manholes, just small concrete inspection boxes that the community could build and manage. These schemes were offered to communities by the UEBD using the same partnership-and-mobilization method. In Colonia Smith *barrio*, where installation began in 1998, a volunteer worker, Ada Victoria, describes how keen they were to have the sewer:

> *Our latrines got full of water and stank. The groundwater here is high, our houses are small and they would overflow in the rainy season. When you tried to dig a new pit the same thing happened. It was impossible to sit down, you were consumed by mosquitoes.*

As part of the earlier water and sanitation project, people in Colonia Smith had received health education and were aware of the connection between insects and dengue fever, malaria and other diseases. Undoubtedly, among the women, this was an additional motivation for building the sewer. The community – with 214 households – managed to raise 400,000 lempiras (US$21,000) for the pipes and fittings. Committees for each sector managed the budget and worked out how to distribute payments and work-in-lieu between better-off and less-well-off households. Ada Victoria describes the time that the project team came as a happy one, 'a time of revolution', when they all worked together up to their waists in mud to install the sewer system. Today, all the *barrio* houses have toilets, and the two health volunteers trained by the project team are still conducting household visits. They insist that each house has a toilet and uses it. If they don't, it represents a risk to the whole community. They are now campaigning to have the stream that runs along the bottom of their community – which is full of rubbish – covered, and to have the municipality bring a regular garbage collection truck. Their motivation, organization and community pride continue to pay off.

Such success stories tend to mask the long and determined efforts some community members have to make to enable such schemes to work. The struggle to bring in water, latrines and, later, sewerage to Colonia Smith – at the same time as other amenities such as electricity, a kindergarten and a health centre – took around ten years. In some other communities, where, due to topography and longer distances, the expense of installation was much higher on a per household basis than in Colonia Smith, communities have had to struggle even harder, in some cases landing themselves with debts it will take many years to pay off. In one community on the sides of a valley on the other side of town, Carbon 2, the community council refused to get involved on account of the obvious expense. A woman householder, Olga Reyes, has virtually single-handedly brought the sewer to their doorstep after pushing the community – with help from her friends at the UEBD – for over six years. And still, the price of the final connection has yet to be set, and they have yet to learn exactly how much money they will have to raise to cover their debt.

In another *barrio*, Soledad, leaders comment that it is hard to force monthly tariffs out of people who cannot afford US$100 to buy a porcelain toilet, even though they earlier contributed their bit to building the sewer. In this vertiginous area, where rooftops descend from the roadside like giant steps, the total cost of the sanitation works for a community of 150 households was over 1 million lempiras (US$53,000). The sewer was finished in 2004, and they hosted a grand inauguration ceremony in the local school. But only 50 households have so far managed to connect. The business of improving sanitation in such *barrios* is a long-drawn-out process – or rather, it never ends. And if it takes years to raise the money to pay back the loans (especially when it is unsafe to hold the money in the community, and when put into an account at the bank, they are obliged to pay an interest rate of 20 per cent), as well as considerable effort to keep awkward residents on the sanitary straight and narrow, local leaders of exceptional commitment are needed.

It seems unfair, from the perspective of an industrialized setting in which water and sewerage connections are laid on automatically to every home, that such a heavy proportion of the costs external to the dwelling should have to be borne by people still near the bottom of the socioeconomic ladder. Other examples in the Americas exist where condominial sewerage costs have been very high – so high, in fact, that if the labour contribution of the *barrio* workers were included, the costs of the system would be higher than those for installing conventional sewerage. With its narrower pipes and more frequent blockages, and the higher risk of breakage because pipes are laid closer to the surface, there are criticisms that poorly managed and constructed, simplified sewerage systems can end up by being a 'poor quality solution for poor people'[44] – despite the sacrifices the *barrio* people may have made. In Tegucigalpa, there have been times when collection of debts from these communities has been abandoned under political pressure. But when that has happened, the Agua para Todos revolving fund stops revolving, the UEBD ceases work on new projects and community sensitization, sanitation in poorer areas is politically sidelined, and the day when the *barrios en desarrollo* will be clean and sewered retreats ever further into the future.

Even given all the potential of low-cost, low-tech, self-build sewerage, and with the excellent motivational and community organization work of a municipal unit such as the UEBD, pared-down sewerage is not an easy solution to the urban sanitation problem on any count – technological, managerial, financial or sociological. Perhaps for this reason, the methodology is still not recognized in standard civil engineering courses in Central America. Biases, fair or foul, remain against it. The situation is not made easier when better-off neighbourhoods are well served with conventional sewers, and their tariffs are relatively tiny in proportion to their means. Subsidy of the better-off as far as sewers and water supplies are concerned is more often the rule in developing country cities than the exception – a reality which people in *barrios en desarrollo* in any part of the world naturally resent. Why should they have to make do with second best? Where there is no option, and the reasons are fully explained and transparent, attitudes such as this can be overcome. The existence of motivated demand and community organization are essential preconditions for launching sanitation projects, but they are still not enough without public investment on a reasonable scale – investment on the scale that is taken for granted by the urban upper and middle classes and by the municipalities that service them. Despite the foresightedness which inspired the establishment of Agua para Todos it is not clear that its mission will ultimately succeed unless costs to the *barrios* can come down.

At least in one important context, the people of Tegucigalpa are gaining sanitary ground. As the city has grown, and the discharged volume of greywater and blackwater with it, the River Choluteca that flows through the city's central valley has become so full of pathogenic and stinking effluvia that, like London's Thames in 1858, it has become unbearable in the summer months. As a consequence, in the wake of Hurricane Mitch and the unbelievable mess that it caused, two major donors – the European Union and the Italian government – arrived offering not only to improve the sewerage system in many of the *barrios*, but to treat the content of the collectors before they discharged their filth into the over-burdened river. The EU's Projecto de Reconstruccion Regionale de America Centrale (PRRAC) has chosen what they describe as a ground-breaking low-cost system of treatment. Its core is that 80 per cent of the contamination in the sewage is eliminated by natural processes connected with the heat of the sun, without injection of oxygen, reducing energy and machinery costs. The plant is able to function on the biogas it generates, and the surplus can be sold. And the 'Great Stink' of the Choluteca during the summer has already been reduced.

As towns and cities all over the developing world lay ever larger networks of sewerage containing ever growing volumes of contaminated fluid, treatment of sewage is becoming an increasingly important issue. In Brazil and Mexico, for example, less than a fifth of wastewater is treated,[45] and the volume of sewage discharged untreated into waterways is even higher in other, poorer, Latin American countries. In India, too, only 13.5 per cent of sewage is treated before being discharged into rivers.[46] And the record elsewhere is not much better: in the Philippines, the vast majority of toilet facilities are privately installed and depend on septic tanks, but none of these are regulated and many are poorly designed, discharging their product directly into stormwater drains, waterways and streets.[47] Sanitation in the form of the flush toilet that everyone prefers is itself a huge potential contributor to environmental pollution, as the Victorians observed. Rivers full of human waste themselves become dilute sewage; they lose their dirt-absorptive capacity and are no longer able to support aquatic life. Cost-conscious and energy-conscious technological options for treating or reusing human wastes from low-income communities of the developing world, rural and urban, constitute another public health engineering frontier.

Having overcome many obstacles and difficulties, Ada Victoria, Olga Reyes and other determined women of the Tegucigalpa *barrios* have their toilets. Even if they have to save water in cans for the flush because the water supply on their hillsides is erratic and the pressure never amounts to much, they do now have decent sanitation. But as we have seen, this is not and cannot be the case for the vast majority of residents in slums and shanties. All the types of sanitation which qualify as 'improved' according to WHO and UNICEF definitions and are lower on the 'sanitation ladder' than simplified sewerage are 'on-site'. This means that the excreta are not removed upon use of the toilet from the household or toilet area. The seat, toilet bowl or squatting plate – since not everyone wants to sit down to perform their bodily functions – is situated over a pit. In the more sophisticated case, a pipe leads the muck away to a septic tank: 50 per cent of sanitation in Japan, for example, is on-site rather than sewered, showing that this does not have to be an inferior technology. On-site toilets in poor neighbourhoods are usually variations of what are known as 'pit latrines'; but since this denotes something markedly inferior to a toilet (or bathroom, rest-room, convenience or other euphemism), the term 'latrine' is avoided in this book. Armies, campers, scout corps and refugee camp inhabitants may dig latrines, but householders with any kind of improved facility have toilets. The idea of an 'improved latrine' is not very compelling to those with a desire for social improvement.

Since so many of the world's low-income inhabitants – urban and rural – are going to have to depend on pit toilets for the foreseeable future, on-site sanitation has been allocated a chapter of its own (Chapter 4). But particular features concerning its use in urban communities are worth noting. Some have already wandered into the picture: the illegality of pit toilets in urban Zimbabwe and the difficulty in Colonia Smith of finding space to build another pit when the existing one is full. Indeed, the problem of digging even the first pit may be acute when living space is extremely congested or the groundwater table very high. In peri-urban Dakar, for example, where the town has grown outwards and absorbed the villages in its path without bringing sanitation, drainage or rubbish collection, the filth of the town is liable to overwhelm the living environment. Ouakam, a neighbourhood of 75,000 people (around 10,000 families), is typically labyrinthine. There is barely room to squeeze between rooms (houses may not be integral structures), doorways, washing spaces and courtyards, and no alley-way is straight or wide enough for a donkey and cart to pass, let alone a mechanized vehicle. This state of affairs is common in many developing world settlements which do not yet correspond to the received idea of 'town' but are no longer rural either.

Where plots are legal and tenancy regularized, as is the case in peri-urban Dakar as well as elsewhere in rapidly expanding urban Africa, the first priority of owner-occupiers is usually dwelling space for their growing extended families. In some cases, as in the old sites-and-services programmes in cities such as Nairobi or Lusaka, people with title deeds may make income out of filling the plot with makeshift rooms, whatever the regulations about density. Where there are newly designated settlement areas, a certain provision of toilets, showers, drains and washing places may be mandatory, but where dwellings are informal, neither water or sanitation will be laid on. Sometimes several households agree to share facilities. But these may become a source of friction if their different members, including children, do not exercise discipline in keeping them clean. In addition, managing their use properly may not be a priority until the family plot becomes really crowded or is no longer next to wasteland where people can go to relieve themselves in the time-honoured way. By that stage, any original facility may be wrecked or overused and overflowing. Space to build new amenities, as in Ouakam in Dakar, may by this point be very constrained.

In the last five years, a low-cost community sanitation programme has been underway in urban fringe communities in Dakar, implemented by the public works agency AGETIP (Agence d'Exécution des Travaux d'Intérêt Public), mainly with World Bank support. Among other targets, such as schools and community installations, the programme envisages 60,000 household installations and is conducting an information and education programme to change personal behaviour. In Ouakam, as elsewhere, the work on the ground is carried out by a local community-based organization, with one team of animators to explain the programme to householders and generate demand, and another team of technical personnel to install the 'works': toilets, showers and domestic wastewater systems. The coordinator of the programme in Ouakam is local NGO leader Birane Ndaye, whose professional life has been committed over several years to transforming his neighbourhood in different ways.

So tightly packed are the households that finding spaces for toilets, showers or soakpits is a serious headache. The technical team have had to wriggle pipes under living-rooms and around kitchens and other people's tiny yards, and householders have had to dig pits in cramped open-air corners. Since Ouakam's inhabitants want pour-flush toilets, and emptying is almost impossible unless a house is on the road, many families have had to find spaces for two toilet pits to use interchangeably. But demand has been high. The costs of installations are subsidized, with a fifth provided by the families, and the motivators with their frequent visits and meetings have proved persuasive, especially concerning the economies to be made in wastewater disposal. Over 50 per cent of Ouakam's families have so far enlisted in the project. Some have gone into debt; the very poor have been given extra assistance. Within a year, the whole area has been hygienically transformed and new community pride has developed. And when the project is over, standards are most unlikely to revert, especially since the necessary technical and administrative skills have been implanted, and Ndaye's community organization will still be active in this and other social improvement areas.[48]

Programmes of this kind are gradually increasing, but are not nearly so common as they should be. In many towns and cities, the inhabitants of slum neighbourhoods and informal settlements are obliged to rely on facilities they install themselves. In Dar-es-Salaam, for example, only 8 per cent of the population is served by the sewer network, while 90 per cent of city households rely on pit toilets and septic tanks.[49] Studies from Zambia, Zimbabwe and South Africa similarly show that, with a few exceptions, most inhabitants of slum settlements rely on simple (or unimproved) pits – whether or not these are legal under local housing law.[50] In Zambia, between 83 per cent and 98 per cent of households in informal settlements were found to be using such facilities. Where there was no toilet at all, typical reasons were that the original pit was full, or there was no space, or the landlord would do nothing because tenancy was insecure or the building illegal. This group of studies shows that the idea of there being little demand for sanitation in poor urban areas is nonsense. In fact, much of the gain in 'improved sanitation' coverage over the past decade has been due to the voluntary installation of better toilets by people in poor and crowded neighbourhoods who have dug their own pits and purchased their toilet bowls or squatting plates out of their own pockets. Just to take one example, private action via the marketplace by householders accounted for all of the gain in basic sanitation coverage in Kampala, Uganda, during the 1990s.[51] Proof though this is of strong demand, it should not be taken to mean that this kind of private consumer action – or self-help as it might be characterized – is leading to an adequate sanitary situation in most urban areas: far from it.

For many of those who depend on pit toilets, the issue of what to do about the contents when they fill up is a major concern – it bothered 75 per cent of respondents in the Southern African studies. Where private services do exist, they may be exorbitant: for example, in Durban, the cost of emptying a pit privately was US$123. The Durban Metropolitan Council did have a subsidized service costing only US$4.50 per pit, but this was only ever extended to the illegal informal settlements during a crisis – a cholera outbreak, for example.[52] An obvious improvement which could be made to the provision of sanitation in cities and towns all over the developing world would be to provide publicly funded or subsidized pit emptying services. Too often, sanitation projects have consisted of toilet construction programmes with carrots and sticks offered to entice 'participation' in the construction phase, but with little consideration given to what is going to happen to the shit after the pit has filled up. Where the issue has been addressed, as in Ouakam, it has been primarily addressed as a design and technical issue – dig the pit deeper or make two pits to be used in sequence. But this usually adds considerably to the expense and to the difficulty of finding space for installation. In Dakar, a lively entrepreneurial business has already been made out of pit-emptying services for middle-class families, but poor households will not be able to afford the monthly US$30 fee, even where the house is near enough to the road for the pit-emptying tanker's pumping equipment to reach it, which is not often the case.

Given that up to 70 per cent of populations in cities in sub-Saharan African countries and in countless towns and cities in South Asia and elsewhere are going to have to rely for the foreseeable future on pit toilets, then incorporating regular pit-emptying into public service provision is an essential step. Many towns have their own informal providers; the scavengers with their night-soil carts who used to be ubiquitous in the Indian sub-continent and in West Africa, some of whom still ply their 'untouchable' trade today, are an obvious example (see Chapter 6). In Dar-es-Salaam, Tanzania, people known as *vyura*, or frogmen, are today's equivalents, and they have their counterparts in major slums in Nairobi, notably Kibera. Emptying compacted faecal sludge out of pits by hand, bucket and shovel is the only possible method in alleyways where no machine can enter or where there is simply no space to dig a second, or alternative, pit. For reasons that can only be attributed to the extraordinary neglect suffered by sanitation during the entire development era, it took until the mid-1990s for the issue of pit toilet emptying to be given any public policy attention at all, and even then it was limited to a level of minimum, and almost exclusively NGO, concern.[53] Today it is still only on the very outer fringes of municipal or donor interest.

An apparently promising approach for areas where lanes and alleyways are too narrow for vehicles is the development of small devices which can empty pits by vacuum extraction without the need for large trucks. In Kenya, a cart-borne pump driven by a small petrol engine known as a 'vacutug' was developed in 1995 by an Irish enthusiast, Manus Coffey, and taken up by UN-Habitat and KWAHO, a Kenyan NGO working in Kibera township.[54] Its expense and other factors means that it is still at the teething stage ten years later. But with the appropriate interest and investment, small-scale service industries could develop around this or similar emptying mechanisms in the future. (The question of occupations and livings made from muck is covered in Chapter 6.) But the fact that it took decades of mounting urban squalor to get such approaches to the starting block seems astonishing. In the 1980s, when engineers were exploring handpumps and rainwater-harvesting devices, and designs were deliberately developed to make community maintenance and repair practicable, the equivalent in sanitation was ignored. Yet without follow-up and maintenance services for the removal, storage and treatment of faecal sludge – services implicit in the provision of water-borne sewerage – the construction of pit toilets in high population density areas cannot lead to 'sustainable' sanitation, and the 'coverage' figures will remain a mirage.

This question returns us to the larger question of the neglect of poor urban areas generally. Part of the reason for this was the assumption that, as far as urban development was concerned, the market and private capital would provide the necessary momentum and international donors need not become concerned. In the early 1990s, as cities in many parts of the developing world began to demonstrate critical levels of environmental degradation, the poor management of municipal space and the dilapidation of infrastructure and serious lack of amenities gradually came to light. Town and city authorities faced with rapidly mounting populations were experiencing overloads on water resources and systems of waste disposal and widespread pollution of rivers and streams. With a long list of management deficiencies and political pressures to overcome, public utilities were fighting a losing battle to provide a functioning service in the face of increasing demand. A World Bank review of the time found that they were locked in a vicious circle. The water and sanitation services they provided were so poor that they could not recover their costs from users and the income generated so low that the services could not be improved.[55]

At the same time, the question of water scarcity and the sustainability of this vital natural resource was becoming a major international issue. First in 1990, and again at the first Earth Summit in 1992, the international water and sanitation community laid down as a principle that water was an economic good to which a realistic price tag should be attached. Boosted by the prevailing ideology of market supremacy, the idea developed that this would happen naturally if inefficient and politically manipulated public utilities were replaced by private companies and subsidization of service delivery ceased. Private providers, with their superior efficiency and responsiveness to demand, would also breathe new life into languishing service spread and manage to penetrate the parts of towns and cities that public utilities had failed to reach. IMF and World Bank structural adjustment packages in the 1990s and early 2000s demanded as a condition of loans the privatization of municipal water and sanitation boards. Immediately, a bitter 'public vs. private' debate broke out. On the one hand, outrage was expressed at the notion of commoditizing the rain and filling the pockets of multinational corporations keen to cash in. On the other, drastic reform was needed to reduce the wastefulness and corruption of many public utilities and to end the invidious practice of subsidizing services to the better-off while providing no service at all to poorer citizens.

In all the clamour that accompanied this debate, one of the most extraordinary features was that sanitation was almost entirely ignored. The polemic – for and against privatization of services, for and against the idea that private capital, business efficiency and demand-responsiveness would increase the spread of services and help attain coverage goals – referred to 'water supply and sanitation services', but the meat of every case referred to water supplies only. The pipes, taps, spigots and drains which would be installed – or not – courtesy of the market, or of the publicly owned and operated public health engineering department, were only connected to water, never to shit or pee. It was the market price-tag on water – admittedly essential for life in a way no toilet can claim – which caused real angst to the anti-corporateers in the first place. And when it came to protestations about people's willingness or ability to pay, or arguments about how to operate redistributive tariffs, eliminate political pressure or extend pipes to under-served areas in lower-income areas, sanitation evaporated as if it had never been there.

In fact, sanitation wasn't ever in the picture, at least not as far as poorer neighbourhoods were concerned. The only kind of sewerage normally installed either by public utilities or by commercial water companies is the conventional kind, and the costs of this kind of sewerage being what they are, its role in meeting the sanitation needs of low-income communities is non-existent. This is the case in towns – the only settings where either type of operator could expect to collect a profit-making revenue from customers – as well as rural areas. The role of international capital and private corporations, and even of most public utilities as currently constructed, in meeting the MDG for sanitation will be negligible. On the other hand, if what is meant by the 'private sector' includes NGOs, small-scale or informal providers, or operators such as Sulabh International, all of which depend on consumer demand as a key operating principle, then its role will be significant. Indeed without such players there would be very few people in the business of low-cost urban sanitary provision in developing countries at all. But that is not what is usually meant by market forces and the involvement of private capital; in international policy debates this has invariably meant the takeover of public utilities by the corporate water companies and their local subsidiaries.

Looking back on the debate – on which a truce has now been arrived at, mainly because privatization did not yield the benefits anticipated and no-one now claims that a change in utility ownership can solve the many problems involved – it is clear that, if the needs of inhabitants of urban slums and shanty-towns for sanitation as well as for water had been factored into the cost–benefit analyses, the case in favour of privatizing utilities would have fallen apart. The private vs. public debate was only ever about water; mentions of sanitation were purely cosmetic. And while sanitation and hygiene require water, the provision of water does not necessarily do anything at all for sanitation – showing that sanitation should lead the debate, not follow along in its wake. The whole debate about public vs. private illustrates how difficult it is to get any serious focus on sanitation when 'water' is the umbrella to whose penumbra sanitation is relegated. Some switch goes off in the brain, allowing the mind to slide neatly past the fact that utilities and water companies are almost invariably also responsible for shit. Where there is no possibility of sewerage, that should not mean that the public health responsibility for the removal of human waste by a service of some kind ends – especially since sanitation is more important for disease control than water supplies. Otherwise, the implication is that public health as an argument for the extension of water supply and sanitation services is picked up and put down on the basis of whim.

The virtual abandonment of poor urban populations to solve their excreta disposal problems unaided, when they not only feel acute need but express demand for decent facilities when they are asked, represents a disgrace on the part of the municipal and development escutcheon. It is a disgrace which would have been incomprehensible to an earlier generation of municipal and sanitary reformers. Urban public health does not conjure the interest, resources or commitment that it did in earlier times. Part of the unfortunate reason seems to be that the 'threat from below' in terms of epidemics of cholera or other life-threatening disease has been reduced by modern medicine.[56] And in spite of the horrendous violent crime rates in cities such as Sao Paulo, Johannesburg and Guatemala City, politically organized violence by the urban poor does not constitute the threat to the established authorities that it did in 19th-century Europe. Maybe in due course, stinking rivers such as the Choluteca in Tegucigalpa, environmental pollution, resurgent disease epidemics, alcoholism and drug abuse, violent crime, increased social distress, and the alienation of tourist visitors and corporate dollars from places without due respect for human dignity or urban wellbeing will turn this situation around.

3

In Dignity and Health

Previous page: A girl enjoys the new toilet at Tulung Elementary School in an earthquake-ravaged area of the island of Java, Indonesia. UNICEF is working with the local NGO Yogyakarta Community Foundation (YKY) to build toilets, drinking water and washing facilities in neighbourhoods and schools.

Source: UNICEF/Josh Estey

During the 1970s, the international community began to pay much more attention to issues of public health in the developing world, and diseases of dirt and squalor inevitably came to the fore. In 1977, when the UN Conference on Water was held at Mar del Plata, Argentina, a determined core of sanitary reformers prevailed on the delegates to declare 1981–1990 the International Drinking Water Supply and Sanitation Decade (IDWSSD). They also managed to obtain international commitment to a Decade target – 'water and sanitation for all' – even though the key instigators privately knew that this was unattainable in the timeframe. At least, they felt, effort would be galvanized. The fact that, 30 years later, the world is still far from reaching the target is an indication less of failure than of how naïve many policymakers then were. In spite of its over-ambition, however, the Decade did manage to launch a new international crusade on behalf of public health engineering as a key to disease reduction. It led to a radical overhaul of precepts and strategies in both water and sanitation – an overhaul that, in the minds of those who worked behind the scenes at Mar del Plata, was badly overdue.

Immediately, drinking water supply – or water in general – hijacked the centre ground. Although the Decade was also supposed to be a *sanitation* decade, it was always referred to as the Water Decade, and drinking water supply is where all the investment and much of the technological attention went. Nevertheless, the Decade did shine a spotlight on the lack of both kinds of service in poor, especially rural, areas, and – importantly for sanitation – it also led to an important reckoning about what needed to be done. It did less well in terms of getting there, however: in 1991, at the Decade's end, there were 300 million more people without sanitation than at its beginning. This loss of ground, due mainly to the spread of amenities failing to keep pace with population growth, was especially pronounced among rural communities and prompted new efforts for rural populations by certain organizations, including UNICEF, the UNDP/World Bank Water and Sanitation Program (WSP), and Dutch and Scandinavian donors. By the early 1990s, the UK's WaterAid – an NGO which came into existence during and because of the Decade – was also paying more attention to sanitation.

At the launch of the Decade and during it, health campaigners constantly repeated the refrain that 80 per cent of sickness in the world was 'water-related'. The term is easily confused with 'water-borne', with which it is often used interchangeably.[1] The implication that the overwhelming majority of sickness is water-borne or is caused by faulty drinking water still pervades popular lore today. The reality, however, is that a much higher proportion of this disease burden is to do with poor excreta control and lack of hygiene. Because water is a highly efficient conduit of pathogenic particles from human excreta, if these find their way into the drinking water supply – as they did at the famous London Broad Street pump in 1854 and still do in many areas of the world today – then the drinking water supply will indeed be an efficient spreader of diseases of the diarrhoeal persuasion. Disinfection or protection of the water supply will help reduce transmission. But poor *sanitation* was and is the much more important underlying cause of diarrhoeal disease transmission, in 19th-century Europe as in Madagascar, Bangladesh, Bolivia, Senegal and everywhere else today. The long ascendancy of the term 'water-related' to describe illnesses more directly related to dirt and squalor has helped to reinforce the obscurity surrounding sanitation. As recently as 2002, WHO's World Health Report seemed determined to empha-size the safety of water in the household, rather than the presence of toilets and hygiene knowledge, as the key to reduction of faeces-related disease.[2]

Some particular scourges – guinea worm in drinking water, for example – are genuinely 'water-related'. Others, such as malaria, have an equally genuine connection to water, because of mosquitoes' and other insects' affection for standing water as a breeding ground. But for the diarrhoeas, pathogens arrive via 'faecal–oral transmission' – meaning that minute infectious particles from someone's shit end up in the victim's mouth and digestive tract. There are many other possible routes apart from imbibing these in water, including hands that have been in contact with unmentionable parts of the body, or with other people's dirty hands, soiled clothes, including children's nappies and underwear, and door handles, taps, pots and drinking vessels unintentionally touched with faecal matter. Small children who play in the dust and dirt of the compound and constantly put their fingers in their mouths are particularly at risk. Many mothers do not realize that children's faeces are particularly full of pathogenic material. They are less concerned about where toddlers 'go' and how the output is disposed of than with older girls and adolescents, whose modesty is more precious. In fact, in many societies they expect their small children to squat on the ground close to the house and discourage them from using the toilet because they see it as risky.[3]

Only in the 1990s did the protagonists of public health begin to challenge openly the idea that 'safe' drinking water was *the* way to reduce diseases of dirt and begin to talk up the need to break the faecal–oral route of diarrhoeal disease transmission. Even now, the misconception that water is the key to disease control lingers on outside the close circle of sanitary enthusiasts. When the problems of unsafe water and epidemic disease are mentioned in news reports or charitable appeals on behalf of inhabitants of the developing world, the real culprits – shit and the lack of sanitation – fail to get an airing. Not only has this prevented sanitation from receiving its due share of popular or official attention, but it has also backfired in terms of water supplies. Investments in safe drinking water were promoted and justified by their notional impacts on health. When studies began to show that few such benefits actually resulted,[4] some organizations backed off from providing poor communities with drinking water supplies on the basis that this was not an important means of promoting child survival. Medical interventions such as immunization were taken up instead because they would have a more potent impact on the toll of child mortality and disease. The fact that nothing, including life itself, and certainly not hygienic practices such as washing of hands, food and kitchen utensils, can go forward without a water supply was, for a while, eclipsed.

In the early 1990s, when public health engineers found themselves downgraded as players in child survival, leading experts subjected water and sanitation interventions to intensive scientific investigation and fought an effective rearguard action on behalf of their central role in disease reduction.[5] This proved difficult. It is hard to measure the impact of a household toilet, children's potty or domestic water tap on a family's state of health, because their acquisition is often a product of wealth and status. Since families that have such items are those for whom ill-health is less likely, the presence in the home of a television set or tiled roof might also correlate statistically with reduction of diarrhoeal disease, thoroughly confusing the issue. However, experts such as Richard Feachem and Sandy Cairncross at the London School of Tropical Hygiene and Health, and many other colleagues in the international sanitary and public health community, were determined to overcome the difficulties of proving the health benefits of 'watsan' (water and sanitation) interventions convincingly. It simply could not be the case that public health interventions were incapable of making any impression on people's – especially children's – health.

Analysis of data from many countries and different types of programme – the impacts of water alone, water quantity as opposed to quality, quality alone, sanitation alone, water with sanitation, and both with hygiene education on numbers of diarrhoeal bouts, other infections, and children's height or weight for age – were undertaken to establish which interventions yielded maximum results and of what kind. No tidy formula has since been universally agreed. But the evidence is incontrovertible that water and sanitation have important impacts on child survival and wellbeing and that – leaving hygiene to one side – sanitation, in the form of safe disposal of excreta, has more effect on reducing disease than provision of water supplies. For example, in the case of diarrhoeal disease, the installation of toilets of the 'improved' variety has been shown to reduce infections by an average of 32 per cent, whereas an improved water supply does so only by 6 per cent[6] (this rises to 25 per cent if cholera outbreaks are included, but cholera would not get into the water supply if sanitation was sufficiently 'improved' to be effective). In 1993, WHO experts set out to rank water, hygiene and sanitation interventions in order of priority according to health benefits. Safe excreta disposal came first, hygiene second and provision of clean drinking water third.[7] This version of public health affairs remains the received wisdom, even if it is not popularly known.

The reason for sanitation's pole position as the main weapon of public health is very well expressed in a 50-year-old visual configuration of the problem – the 'F-diagram' (Figure 3.1).[8] The only way to put in place an effective barrier between the disease-causing agents present in faeces and all the pathways they use to attack human health is to confine the faeces. Water, both in terms of its quality or 'safety' to drink and in terms of its plentiful supply for use in personal hygiene, only comes into the picture further down the line (although there is a strong case to be made for its primary use in washing fingers immediately after defecation). Shutting up, removing or channelling away the shit out of harm's way is by far the most important way of removing the primary source of infec-

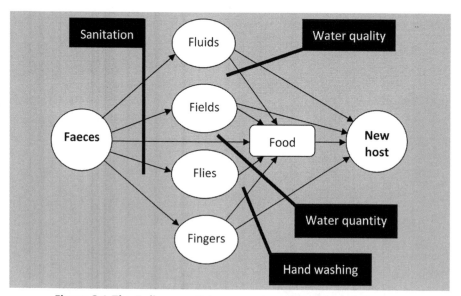

Figure 3.1 *The F-diagram: Primary routes of faecal–oral diseases*

Source: E. G. Wagner and J. N. Lanoix (1958) *Excreta Disposal for Rural Areas and Small Communities*, WHO monograph series No 39, WHO, Geneva

tion, in connection not only with pathogens and parasites present in faeces, but also with insects such as mosquitoes, flies and cockroaches that carry germs out of the smallest deposit of excrement, or from a midden, dung-heap or pit, into the environment, around living areas and onto food.

In the triumvirate of water, sanitation and hygiene, however indivisible one might want them to be, sanitation is therefore the *primus inter pares*. Interestingly, the supremacy of sanitation in controlling disease was also the finding of a recent poll carried out by the British Medical Journal among 11,000 of its contacts around the world. Offered a variety of contenders, including immunization, they resoundingly and sensibly voted the toilet the greatest medical milestone of the last 150 years.[9] Unfortunately, it seems that politicians, celebrities and philanthropic corporate donors are not represented in these pro-sanitation ranks, being willing to couple their names only to delightful water, rarely to nasty shit. Water is certainly needed for hygiene. Good hygiene, especially hand-washing, can prevent the transmission of diarrhoeal disease: evidence shows that diarrhoeal bouts may be reduced by 45 per cent by washing hands with soap after defecation and before eating,[10] and protection of household drinking water against faecal contamination is also key. But there is no substitute for a sanitary toilet. Other interventions are secondary. This is not to diminish their importance: all methods of disease reduction should be exploited. But though some public health professionals appear eager to find one – to save costs, to duck the difficulties of promoting sanitation or simply to avoid such an indelicate subject – there is no short cut. The confinement and removal of excreta, at least until it is sanitized and safe, is a public health must.

Unfortunately, that does not make sanitation a must with all public health authorities or with potential customers. For many years, the over-emphasis on water as the principal driver of public health, and the much higher demand from customers for water supplies as compared to toilets, has skewed interest and investment. In fact, even in the case of water, consumers expressing demand were not thinking of health but of convenience and livelihood. Where programmes had a hygiene education element, often lectures on germs and worms too tiny to be seen did not do a great deal initially to generate enthusiasm for toilets – at least in settings where there was no other impulse for demand. Conveying health and hygiene information convincingly, so convincingly that people change well-worn and intimate habits, is a tough assignment and takes time.

With no more than a superficial effort to challenge forms of sanitary behaviour entrenched over generations – whose characteristics few researchers bothered to examine – and in the face of superstitious or non-scientific theories of disease, those sanitation programmes that existed had a difficult time. As a result, for many years an assumption was made that 'demand' for toilets in poorer parts of the developing world, and especially in rural areas, did not and could not exist. They were culturally opposed, for one thing. And they were costly. People could not want to remedy, and expensively, an unperceived threat to their own and their children's survival. So those public health engineers who were concerned about the issue believed that the only thing to do was to install toilets in homes and institutions at donor or public expense and hope that over time people would get the idea. An earlier generation of public health reformers exemplified by figures such as Edwin Chadwick and Joseph Bazalgette had managed to inflict sanitation on entire populations. Perhaps the exercise could be repeated?

The problem was that it could not. Not anyway in rural areas, where today 2 billion people still lack access to 'improved' sanitation. One reason is that, where sewerage is not the means of excreta removal, sanitation is not possible by executive high command. A sewer is a communal asset – as is a piped water system – and can be centrally managed. But where a local sewer is impracticable, sanitation is a matter of separate household installations. Public works departments have sometimes tried to impose sanitation, by arriving in a village and building pit toilets in everyone's backyard whether they wanted them or not, or by issuing a proclamation that people must build toilets under threat of a fine or worse. But this approach usually fails: if people don't want a toilet and its value is not explained to them, they don't tend to use it (or they use the structure for something else), and it soon falls into disrepair. In Madagascar, in response to a cholera epidemic in 2001, threats and menaces, not educative explanation and information on health hazards, were used to force people to build toilets, and as a result most of the facilities were never used.[11]

Unless there are multiple connections to a pipe leading to a septic tank, on-site sanitation has to consist of discrete installations. In the on-site scenario, each household builds and operates their own bathing and toilet facility – as they organize their own rubbish disposal, water filtration or disinfection, and waste-water drainage. There is no central system of taps or levers for the sanitary high priests to administer on behalf of the group, and it is difficult to admonish those guilty of hygienic misdemeanour by legal action or service cut-off, thus bringing them into line. Housing regulations in much of the developing world do not require the installation of bathroom or toilet. Nor, until very recently, have local councils or 'water and sanitation committees' set up to manage and maintain new water and sanitation infrastructure presumed to suggest that every household in a neighbourhood toe the local sanitary line. Only this kind of enlightened local governance around water and sanitation binds households with one another – and it is never an easy task to do this, as the examples of simplified sewerage systems from Tegucigalpa, Honduras, in Chapter 2 demonstrated. Household-based systems require a high level of community organization not needed in neighbourhoods whose sewers, drains and rubbish disposal systems are operated centrally, typically from public funds by public servants and public health engineers.

For around 20 years, most rural sanitation programmes consisted of building 'latrines' – not only in communal facilities such as health centres and schools, but also in households, free of cost or at heavily subsidized rates. The hinterland in countries such as Nigeria, Senegal, India, Pakistan and Nicaragua became studded with a new kind of monument to development folly: solidly constructed temples of convenience, costing twice as much as and built of stronger materials than the dwellings in which village people lived. The basic premise of many schemes was that the 'demonstration' toilets installed in the compounds of chiefs, councillors, local schoolteachers and other figures of standing, together with the training of local masons in their construction, would eventually lead to sanitation spread. What tended to happen, however, was that several years later, the same handful of toilets existed, perhaps well used, perhaps not, but looking rather more dilapidated than before. Little toilet 'take-up' could be seen elsewhere, and the local masons were making a living out of house construction as per usual.

Thirty years have passed since the goal of 'sanitation for all' was first set out at Mar del Plata. Many lessons have since been learned about how to take forward the new public health crusade among people whose own resources barely provide them with a decent roof over their heads, let alone with the means for sanitary home fixtures, in environments where excreta disposal infrastructure of any kind was previously unknown. And still today, the difficulties facing this new crusade are legion, and not nearly as well understood as they should be. Apart from the need to attach more importance to sanitation, there are still wide gaps between the perception of public health requirements and the realities of private consumer practicalities and tastes. Standard nostrums still need to be challenged – for example that there is no latent demand for sanitation among poor inhabitants of the developing world and that people whose custom is 'open defecation' have no appreciation of clean and healthy living. Every aspect of conventional wisdom in the vexed area of personal hygienic life needs to be opened up to local examination and appraisal for programmes to work. This requires a concerted attempt to dispel the taboos unnecessarily hindering progress.

Although sanitation did not make it into the 1980s 'child survival' portfolio, the threats to child health and survival posed by excreta are stark. Raw faeces, even that of healthy people and especially of children, constitutes a staggering danger to health. Just a small splodge can contain high concentrations of viruses, several groups of potentially pathogenic bacteria, harmful species of protozoa, and a generous helping of helminths or parasitic worms.[12] Given their daily proximity, the risks to human health from faeces are far higher than those from any other substance. Fortunately, natural design – look and smell – makes them so unpleasant that we instinctively shun them. But disgust, coupled with taboo, is not an adequate disease-avoidance strategy. Knowledge too, and the conversion of knowledge into changed behaviour, is essential. A large proportion of the world's citizens have yet to be scientifically acquainted with the lethal risks of faecal contact and how to adopt avoidance strategies. These subjects should be taught mandatorily to young people in every pre-school, primary school and non-formal education syllabus throughout the world. Sadly, they are not.

Millions of the viruses and bacteria present in a typical gram of faeces are simply part of what lives in our digestive systems to help make them tick or are the product of symptomless infections forming part of our natural defences. Others, including many of the same bacteria in pathogenic form, are the cause of the many mild and serious types of fever and running stomach billions of instances of which beset adults and children during any given year; only a proportion of these are life-threatening. Some of them can be easily dispelled by a healthy person. Others are much more threatening where health facilities are poor or non-existent, and hygienic knowledge and basic remedies are not available in the home. Thus in poverty-stricken parts of the developing world, a common complaint can become a killer, or at least result in depleted wellbeing, affecting a young child's nutritional status and stunting healthy growth.

Of the over 2 million deaths a year associated with diseases of dirt and squalor, most are due to diarrhoeal disease, and of these the vast majority are in children under five years old, almost all of them in developing countries (Figure 3.2).[13] And the mortality figures only represent the tip of the illness iceberg: tens of millions of children suffer repeated bouts of diarrhoea during their early years. The spread of oral rehydration therapy (ORT), in which UNICEF has been closely involved for the past 30 years, is important in tackling one of diarrhoea's most lethal symptoms: dehydration caused by loss of fluids. But some diarrhoeal infections are not susceptible to ORT – bacillary dysentery, for example, which causes a persistent bloody flow. Such infections require

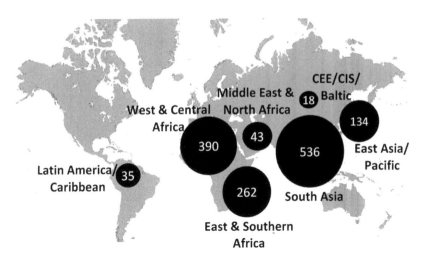

Figure 3.2 *Estimated annual deaths of under-fives from diarrhoea (1000s)*

Source: WHO (2007) *World Health Statistics 2007*, WHO, Geneva

antibiotics or other therapies. There is, in the end, only one way to reduce definitively the toll of childhood diarrhoeal disease and death, and this is to prevent
as many infections as possible in the first place. Sanitation – especially proper
management of young children's faeces – is not only fundamental to keeping
pathogens out of contact with children: once they are sick, lack of faecal
containment is also complicit in the poor quality of their care. Without a toilet
and the means of washing clothes and bedding, a patient with severe diarrhoea
is difficult to nurture effectively.

Diarrhoeas are the most important excreta-related problem, but intestinal
worms, whose eggs mature in faeces and enter the body through the feet or
mouth, also do great damage. The presence of faeces in compounds, pathways,
fields and other places where 'open defecation' is the norm puts adults and
children – especially if they run around barefoot – at risk. There are around
133 million cases worldwide of ascaris (roundworm), trichuris (whipworm) and
hookworm infestation, whose damage is more far-reaching than people realize.
A typical ascaris load diverts around a third of the food a child consumes, and
is also an important cause of asthma. Hookworm is a frequent cause of
anaemia. Trichuris leads to stunting in children and to chronic colitis in toddlers,
a condition which often persists for so long that mothers may think it normal
and fail to seek medical help. Children in poor environments often carry 1000
parasitic worms in their bodies at a time.[14] When treated with de-worming drugs,
they may immediately experience a growth spurt, showing how devastating the

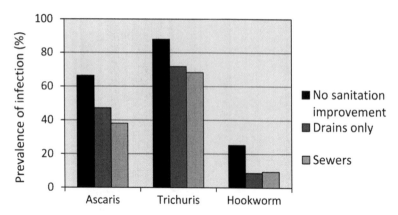

Figure 3.3 *The impact of sanitation on helminth infections, urban Salvador, Brazil*

Source: L. R. S. Moraes, Jacira Azevedo Cancio and Sandy Cairncross (2004) 'Impact of drainage and sewerage on intestinal nematode infections in poor urban areas in Salvador, Brazil', *Transactions of the Royal Society of Tropical Medicine and Hygiene*, vol 98, pp197–204

nutritional impact has been. In some cases of serious worm infestation, children's cognitive development is harmed.[15] Studies show that sanitation makes a significant difference to the parasitic caseload (Figure 3.3).

Also complicit in poor sanitation is trachoma, the second most common cause of blindness worldwide. This is usually described as a 'water-washed' disease because it is caused by dirt and germs getting into children's eyes, and washing the face is an important means of prevention. But flies such as the *Musca sorbens*, which breeds in scattered human faeces, also transmit trachoma and would be deprived of a breeding site if excreta were locked away. Other so-called water-related diseases, such as intestinal bilharzia or schistosomiasis, are also reduced by hygienic sanitation. Because the parasitic worm – or schistosome – completes its life-cycle in the body of a snail, and wading in water containing infected snails is the means of infection, bilharzia is commonly described as water-related. However, the parasite enters the water and thence the snail via the faeces of an infected person and sanitation reduces this possibility. Therefore bilharzia could as well be described as excreta-related. A programme to install household facilities in a particular quarter of the town of Saint Louis, Senegal, succeeded in dramatically reducing cases of bilharzia in the community.[16] It transpired that infected adults had been using a local ditch under cover of night for defecation, and children who waded in it thus became exposed to the parasite-bearing snail. Once people appreciated the connection and toilets were installed, the level of cases dropped.

As public health and nutritional experts re-established the important connections between sanitation and hygiene and child health, some of the most prominent – notably Steven Esrey – began to emphasize health perspectives other than diarrhoea. Esrey suggested that the spectrum of dirt-related diseases had not been given their due credit for causing faltering in children's growth. He proposed that anthropometric measurement (height for age) was a better indicator of the cumulative impact on child health than the number of bouts of diarrhoea a child suffered.[17] For example, a study in the Gambia found that intestinal parasites were present in children 76 per cent of the time, which can be compared to the much lower presence of diarrhoea (14 per cent of children in any one week). This suggested that more mundane excreta-related conditions can have a worse long-term health effect than the more dramatic and frightening attacks represented by fevers with diarrhoeal symptoms. As a result of new research and its publication, the stunting effects of poor sanitation and hygiene on a child's wellbeing became better recognized. So did impacts such as days lost to schooling: just as one example, 3.5 million school days are annually lost to Madagascan children because they are sick from excreta-related disease;[18] worldwide, the numbers run into the hundreds of millions. The term 'DALY' – disability adjusted life years – also came into use by health statisticians at this time, as a nerdy way of measuring the impact of ill-health on a country's economic productivity. Despite efforts by the World Bank and WHO, however, DALY computations have never really caught on as an influence in national development planning or budgetary decision-making.

As in 19th-century Europe, outbreaks of life-threatening infectious diarrhoea provide a more powerful political driver behind sanitary interventions and investments than carefully argued case studies about days lost to work, school or national productivity due to people feeling unwell. Cholera, from which people can die from shock within hours of the first symptoms appearing, still engenders such horror and stigma that health ministries in some countries refuse to acknowledge epidemics, preferring to hide behind terms such as 'acute watery diarrhoea'.[19] And cholera is still very much present in the world today; in fact it is more evident in some regions of Africa and Latin America than it was some decades ago despite the wider spread of primary healthcare services. When cholera arrived in Brazzaville, Congo, in February 2007, doctors had never made its acquaintance before.[20] During Madagascar's terrible 1999–2001 epidemic, there were 35,000 registered cases and 2300 deaths.[21] This prompted the panicked authorities to embark on their first ever campaign to promote family toilet construction. Although this was a breakthrough in the political sense, little was actually achieved because force, not hygiene education, was used. However, the lesson was learned.

What kinds of beliefs or ideas make people shun toilets? And do such ideas merely exemplify ignorance or could they actually have some useful disease-preventive qualities? Anthropological studies suggesting answers to this questions are uncommon, but were carried out in Madagascar in the wake of the cholera epidemic.[22] In parts of the country, according to popular wisdom, a structure to store faecal debris is culturally unthinkable: taboos – or *fady* – forbid its use. This is because one person's excreta should not be put on top of another's – a sensible rule where the heat of the sun is being used to deodorize and sanitize wastes – and bare feet should be kept from coming into contact with detritus left by someone else. Storage of faecal waste below ground is also held to contaminate the dead – which, since ancestor veneration is central to Madagascan culture, is a powerful argument against the digging of pits for muck. Some people object that if grain is stored below ground, this precludes storing excreta in a similar way.

A more general answer is obvious: a place in the open air, with a fresh breeze, and distant seclusion from the eyes of other people has to be a nicer place to use for bodily evacuation than a tiny, hot, dark hut in the middle of the backyard, with the possibility of smells, insects, and unappealing remainders and reminders left by someone else. Indeed in settings where rural people are either far too poor to afford a toilet of their choice or unwilling to consider the possibility, the best approach may be to educate them to choose a place far from local streams and paths, and cover their stools with soil by what is known as the 'cat method'. When the 'sanitation ladder' is shown to people in Laos and elsewhere as part of hygiene education, the 'cat method' is often depicted as the bottom rung.[23] In 1993, when cholera began to pose a serious threat in Mexico, the Ministry of Health advised people to keep a spade near at hand and cover their excrement with lime or earth.[24]

One of the few travel writers to mention sanitary habits is V. S. Naipaul. Writing about India in 1964, Naipaul declared that the Indian peasant suffered claustrophobia if 'he has to use an enclosed latrine'. He also complained that the society as a whole suffered a collective blindness towards the large numbers of people to be seen squatting anywhere and everywhere, by the roadside, on the beach, on river banks, railway lines, paths and open land. As a result his book *An Area of Darkness* was banned in India as offensive. 'Open defecation' was a subject long prevented by Indian squeamishness from being publicly addressed. In a later book about a journey through the lands of Muslim believers, Naipaul records a conversation in Malaysia in which his spiritual informant told him that one should never urinate or defecate near a house, in a place where the result will cause a nuisance or near a stream from which people take water. 'It is a holy teaching and it is applicable in our life. So I took it as something we have to follow.' This interdict was conveyed when the informant was around 12 years old.[25] It is not the case in religious or parental teaching about personal habits that no consideration is given during their upbringing to adolescents' health, dignity or public cleanliness.

Descriptions of pre-industrial sanitation arrangements in the rural developing world are not easy to find: if anthropologists of the colonial period studied these affairs systematically, their legacy has been lost. Unlike customs surrounding sex, procreation and death, dealing with bodily wastes appears to have suffered an anthropological blackout as powerful as any cultural taboo. Occasionally, colonial officers touched upon the subject. In 1940, one wrote that householders in rural Uganda were afraid to use a pit for defecation because its fixed location would give sorcerers easy access to excreta for 'hostile purposes'. The problem was overcome by digging the pits so deep that sorcerers could not reach the faeces.[26] Personal experience or notation by enthusiasts such as John Pickford, one of the earliest stalwarts of low-cost sanitation in the developing world and founder of the Water, Engineering and Development Centre (WEDC) at Loughborough University in the UK, provide odd glimpses into such customs. From these, certain common themes occur. One is the effective use of natural ecological processes for excreta management, including the desiccating powers of sunshine and consumption by scavenger dogs and pigs. In a far less crowded and environmentally polluted world, the systematic use of these methods indicate significant respect for the polluting qualities of excreta – even if the impetus was often religious or ritual, not public health – and their potential harm and nuisance. Another is a fear of insecurity, especially for women, and the need for privacy associated with female modesty.

In Madagascar, as in India, Vietnam and elsewhere, people in low-income communities living on the coast – fisher people, for example – often empty their bowels onto the sea shore, and let the high tide wash the detritus away. Away from the coast and from lakes and streams, people find discrete places in fields or in woodland, just like hikers and bikers in industrialized countries who go camping in the countryside. In Vietnam, a recent study found that people living at the beach much preferred their sand-dunes to any proposed toilet facility. There was even a local saying: 'first the dunes, second the fields'.[27]

In the Qabane Valley in highland Lesotho in the early 1990s, an anthropological study found a quite complicated set of rules about where people 'went'.[28]

Set places for defecation were reserved at a distance from village dwellings and lower on the mountainside, and at night it was difficult to reach them especially for the young, the aged, the disabled and the sick. The study also commented that, despite the custom of depositing excreta on ash heaps outside each household – this is a high, and cold, area where fires are frequently lit – the villages and defecation sites were remarkably free of faeces. Eventually the researchers realized that, as in countless rural communities all over the world, village animals, notably dogs, acted as faecal vacuum cleaners. There was a pecking order whereby a child visiting the ash heap was followed by a waiting dog, whose initial act of consumption led to the arrival of a pig, who in turn gave way to chickens pecking among the fragmentary remains. The rules about what meat is allowed to be eaten in certain religious codes are directly connected to the eating habits of certain small livestock, whose presence in the community is nothing to do with food-raising or food-hunting, let alone enjoyment as household pets, but is deliberately tolerated for a certain unsavoury purpose.

In many societies, 'facilities' for men and women have been traditionally separate, or they may go at different times of day. In the Lesotho study, this was directly connected to ideas of 'respect'. It was embarrassing to disturb someone in the act of defecation at any time, but if the person was a member of the opposite sex it was far more discomforting. In Lesotho, fathers and daughters-in-law retain exceptional distance and mutual respect, and the idea of meeting each other at a defecation site would be appalling. There can be no doubt that these kind of sentiments are common in untoileted societies around the world. The separation of facilities for men and women in every society, including industrialized countries, is a carry-over of the same idea. And not only modesty may be at stake: some societies believe that any mingling of male and female faeces, especially with the addition of menstrual blood, renders people sterile.[29] The spreading of stories such as this was surely a useful way of keeping boys away from girls and men away from women at moments when they might unavoidably be exposing their sexual organs.

In parts of Madagascar, any transgression of the rules for men and women are subject to correction by the local 'wise man' or community head. The maximum fine is the forfeit of a zebu (local cow).[30] In southern Ethiopia, men and women are similarly barred from using the same place as a toilet – with the consequence that, when a household facility is installed, women may be prevented from using it; alternatively, men may refuse to perform their functions in the 'ladies room'. The preservation of toilets for women's exclusive use is more common in societies where they are secluded. Toilet huts are also avoided by children frightened of being shut up alone in the dark, of snakes, smells and sorcery, and of falling into the pit. Not only rules about where to 'go' and when, but also those about what to use for cleansing may be strict. Traditional materials where water is short include stones and corn cobs and other vegetative waste.

In villages in the Rann of Kutchch, western India, a sparsely settled desert area, separate toilet areas at the extremity of the village are assigned to the two different excretory functions.[31] Although these are open to the sky, they have high stone walls to provide their users with privacy. Kutchchis might well argue that their sanitary system meets many of the necessary public health criteria: it confines the material away from dwellings and waterways, and there is plenty of space. Within a few hours the burning heat of the sun has desiccated and hygienized each small offering. In other areas of rural India, a piece of land away from the hamlet may be set aside for sanitary use. The custom is for women to rise before dawn and go there under cover of darkness, carrying a small pot of water for posterior cleansing. An account of living in a Punjabi village by a European woman describes how the mother of her partner would wake her at 5.00am for a companionable joint outing.[32] If she failed to get up, it required great feats of unselfconsciousness to use the area during daylight as it adjoined the local bus-stop. This also exposed her to taunts and ridicule: she lost people's respect and was more vulnerable to sexual or physical harassment as a result.

In the increasingly crowded landscape of the contemporary world, these aspects – personal dignity, female modesty and security from sexual attack – are becoming important reasons why, even in rural areas, traditional systems of attending to bodily needs are ceasing to be personally congenial. In many environments, adolescent girls fear being harassed when they go to the bush for their daily outing. Where distances of several hundred metres have to be walked, and it is only acceptable to do this in the dark – as is the case in a number of environments where women are traditionally sequestered, in Africa as well as Asia – great difficulties for women are entailed. Girls in South Asia have traditionally been taught to submit their interior mechanisms to extraordinary discipline so that they can 'save themselves' the entire day and wait for darkness to descend; to be caught out, seen going to a public facility or having to squat down in a place used by other people, especially men, is to be morally 'loose'. The effect of holding themselves 'in' for so long can lead to women eating or drinking little during the day, which can have an effect on their state of health or internal organs. Women in this predicament are often keen to have a household toilet – if they are not silenced by taboo and feel able to say so. The injunction to purity, modesty and the natural desire for privacy are becoming increasingly difficult to reconcile with the old ways of doing things in environments subject to overcrowding, the presence of strangers and other attributes of demographic change.

In parts of Africa, security can have a different aspect in sanitation practice. There are fears of wild animals or snakes, as well as of attack by unfriendly neighbours. Villages in areas traditionally vulnerable to conflict build their housing closely packed for self-defence. This pattern of settlement is common in West Africa, Ethiopia and Madagascar, as well as in China, North Africa, the Middle East, Pakistan and Afghanistan. Vernacular architecture often reflects fear of hostile intruders, as do boundary walls, entrances and constricted household and village layouts. And so can sanitary systems. In a Nigerian community in the Afikpo local government area, Imo State, a community which came within the reach of a UNICEF water and sanitation scheme in the 1980s, the villagers had their own public sanitation area within the village with separate areas for men and women. The need to avoid being attacked on trips outside the village required that shyness and embarrassment be overcome. Use of the facility was seen in much the same way as a visit to the water tap, as an opportunity for socializing and gossip. Each area had a trench down the middle and logs situated either side as seats. In this setting, conversion to an improved community facility was welcomed.[33]

The case of Afikpo illustrates that understanding of the local situation and attitudes is essential: a few miles away, in communities where houses were scattered, there was virtually no interest in communal toilet facilities, which, when provided, were barely if ever used. Indeed, in some more isolated areas, the foolishness of glorifying excreta by building a house for it was greeted by the local inhabitants with mirth. So annoyed were they that the authorities were trying to impose such a practice – an indication of strong taboos which no-one thought to enquire into – that they refused to construct any building of this kind. In 1986, the chief of one such a community was threatened with imprisonment because his village had failed to comply with the sanitation order. The villagers therefore built three communal-use latrines according to the prescribed design, with doors. They then attached locks to the doors and left the keys in the charge of the chief. When the sanitary inspector visited, he was delighted to find that the latrines were so clean.[34]

Although relatively little is known among sanitary researchers about long-standing traditional attitudes and systems, every society, as well as most religions, has had codes of personal cleanliness. These include rules about where a person should perform their bodily necessities, how to clean the private parts, and how such practices equip a person to be in a state of purity and fitness to perform worship and prayer. In their day, these systems of public health may have functioned adequately, even if they have subsequently broken down where settlement patterns have changed and lifestyles established over centuries are under threat from many different directions. There is far less room for 'empty' spaces in the landscape, vegetative cover is reduced, and in many settings the natural processes of sun, wind and water can no longer bear the absorptive load. However, there may still be places in some sparsely settled rural areas where traditional systems operate effectively. In the hot, dry countries of North Africa, for example, it has been reported that there was less transmission of disease when people used the open fields than when they used unimproved latrines.[35] But the question has very rarely been studied. Effective faecal management has invariably been seen by public health engineers as identical to the spread of the toilet. Yet this may not actually be the only or the best way to look at the problem.

When people are convinced that an existing pattern of behaviour suits them perfectly well, they will take some persuading to adopt one that is imposed from outside, concerns matters which are intimate and even taboo, and requires levels of investment in household improvements not readily to hand. Just as people are only willing to adopt measures of family planning when bearing a large number of children ceases to be a source of wealth and instead becomes a strain, they will only contemplate 'improved' sanitation when some kind of change in ideas and values transforms the concept of a toilet from something despicable and unclean into an asset. Pressure on living space and social aspiration can tip the balance from the old ways to the new, especially where women want to protect their privacy. The importance of having on offer decent, aesthetically pleasing and functional equipment – the satisfaction of consumer taste – should not be underestimated.

Sensitive discussions with local people tend to reveal more willingness to adopt toilets than has typically been attributed to the inhabitants of the rural developing world. But embryonic demand is bound to be based on the need for privacy, modesty, respect, security, social status, disgustingness reduction, environmental cleanliness, protection against sorcery and the other ideas that shaped natural sanitary systems. In this perspective, disease control is way down the list. Perhaps if this had been understood at an earlier stage in the modern sanitary revolution, things might have gone rather better …

One of the earliest large-scale programmes to break the mould of conventional approaches to toiletizing humanity was undertaken by an NGO, the Ramakrishna Mission Lokasiksha Parishad, in West Bengal.[36] In 1990, the Mission proposed to its financial partner, UNICEF, that an 'intensive sanitation' programme be launched in Midnapur, which, with over 8 million people, was India's most populous district. (In 2002, Midnapur was split into two districts, East and West Midnapur). Here, as throughout the Ganges delta, the countryside is water-rich and densely settled. Brilliant green paddy fields are interspersed with clusters of thatched roofs nestling under shady palms; every corner of the fertile landscape is occupied by homes and cultivated patches. In the villages, narrow paths skirt compounds and water-ponds, groves of fruit trees and vegetable allotments. No house is far from the next, often they are separated only by a handkerchief-sized plot. Some of the houses are substantial, but many are very poor, no more than a tiny room with one bed and an outside space for a kitchen. The area may be designated rural, but is more like garden suburbia, imperceptibly taking on semi-urban characteristics as the huge city of Kolkata spreads outwards and swallows its hinterland.

According to the 1991 Census, 'open defecation' was then still practised by over 95 per cent of the rural population of West Bengal. In the rainy season, epidemics of cholera, typhoid and other diarrhoeal diseases were common. The Ramakrishna Mission had been working in deprived parts of the state since the 1970s, and found that handpump drinking water, nutritional supplements and immunization failed to make any impact on the health and survival rate of children under five. The abundance of excreta on paths and in and around people's houses, and the lack of any hygienic system of sanitation, was to blame. The Mission's social workers first began to promote toilets through their networks of youth clubs during the 1980s. At this time, a standard pour-flush, twin-pit toilet had been designated as *the* improved facility for sanitation programmes in rural India; however, this facility cost 2000 rupees (around US$50 at current rate of exchange but significantly more then), a very large sum by rural Indian standards. Even though 60 per cent of the toilet cost was subsidized, progress was slow. Nonetheless, their experience convinced the Mission that if they changed their policy entirely, cancelling the subsidy, offering a cheaper product, and building interest by community mobilization, there could be that magical thing – demand – for sanitation in rural West Bengal. Women especially valued the privacy and round-the-clock availability of a toilet at home, once other hurdles were removed.

The 'intensive sanitation' programme that UNICEF agreed to support in 1991, and to which the state government gave its blessing, started from the premise that demand for sanitation would be created. The first priority would be mobilization and awareness-building; technology and toilet construction came second. The Mission conducted motivational camps and instruction sessions for all kinds of personnel: teams of motivators, village masons, local councillors, youth clubs and women's groups were all subjected to intensive hygiene promotion. To begin with, relatively few toilets were built; but by late 1994, the Mission had reached more than 2600 villages and over 52,000 toilets had been constructed entirely without subsidy. It had proved possible to dislodge age-old habits if promotional messages were repeatedly and persuasively applied. The teams of motivators, most of whom were youth club personnel, found that it took an average of five visits per household for persuasion to pay off. Over time, the cleanliness of the village and the fact that visitors from Kolkata were able to enjoy the use of city-style amenities became important selling points.

Each motivator visited around 100–200 families. They explained about the dangers to health of faecal matter lying on paths and around the village, and they displayed a catalogue of possible toilets. These ranged from 300 rupees (US$7.40) for the simplest polished cement pan with slab to 3000 rupees (US$74) for a twin-pit pour-flush with ceramic pan, cement-ring pit linings and fancy brick cubicle. Potential customers were also offered an interest-free loan if they were willing to put down half the price. Production centres were established, employing local masons to make slabs and pit-lining rings, and women to make pans and water-seals. A new local employment, manufacturing and sales sector was developed around a previously unwanted and unknown consumer item. At any time in the early 2000s, toilet pans and slabs mounted on bicycle rickshaw carts could routinely be seen on the narrow roads of Midnapur, on their way to installation in the houses of new customers (Figure 3.4).

All the time, thanks to the motivators, education about safe handling of drinking water, food hygiene, disease transmission, solid waste disposal and the need to eliminate faecal matter from the environment was being passed on. By 1995, 1000 youth clubs had been mobilized, and nearly half the villages in the district had been reached. After this, the Midnapur district authorities and the local councils began to throw their weight behind 'saturation coverage'. This helped to bring difficult customers – the last 10 or 15 per cent – around. In cases of extreme resistance, pressure would be applied in the form of an off-duty policeman casually visiting the house. For the extremely poor, the no-subsidy rule was relaxed and local resources provided. By 2006, the two Midnapur districts had achieved 100 per cent coverage.[37] By this time, leading NGOs in other districts had been enrolled, and the state had put its own resources behind

Figure 3.4 *Transporting pans and slabs in West Bengal*

Source: UNICEF Kolkata

total sanitation throughout West Bengal. From 12 per cent in 1990, coverage has risen to 40 per cent using social mobilization methods.

The Midnapur story, for all its success, shows that large amounts of time and special investments of effort are needed to get something so radical as an entirely new attitude to household excreta management off the ground. It also shows that when the approach has reached lift-off, and if significant political and financial backing are then forthcoming, the sky is the limit. In Midnapur, the long years of effort by the Ramakrishna Mission, and the partnership and backing of UNICEF, pioneered the way for state and local government commit-ment to developing the strategy and backing its repetition on a state-wide basis. The more sobering aspect has been that the Midnapur experience had not been able to be replicated anywhere in India outside West Bengal. Apart from the unusually energetic and committed performance of NGOs and government, perhaps the crowded landscape and the strong voice of women combined to engender a latent demand which so far other sanitation players in India have found it less easy to discover. There are other successful sanitation initiatives in other Indian states – for example, there are now over 5000 villages around the country which have reached verified 'total sanitation' and been awarded a Presidential Prize for the achievement. But their context and impetus – to which we will later return – is different, and the scale of toilet production and take-up either at district- or state-wide level is so far not comparable.

At the time that the Midnapur effort began, the reversal of emphasis from technology to social mobilization was revolutionary in sanitation programmes. It is still far from universally accepted, but thanks to strenuous promotional work by sanitary practitioners and international donor organizations such as the UNDP/World Bank Water and Sanitation Program (WSP), examples today are much more numerous. Instead of being cutting edge, such ideas – at least among those who are actively seeking to upgrade sanitation among low-income groups and within the international public health community – are beginning to become mainstream, and government departments too are buying in. Today it is becoming more usual to find that a process of community consultation, motivation, persuasion and salesmanship for the clean-village-and-toilets idea has preceded actual construction. Programmes based on social mobilization and marketing are underway in a number of countries, but some are still very young and have many hurdles of local adaptation, acceptance and accelerated spread yet to overcome. Their contribution to child survival is no longer challenged, and although there are many instances where data on programme impacts are not yet available, the importance and value of social mobilization efforts are nowadays taken on trust.

Take the departments of Boaca and Chantales in rural Nicaragua, for example, where a water and sanitation scheme is operated by the state water company, ENACAL. In this hilly area, villages are small and scattered and agricultural productivity poor. The larger plantations have been laying off labour in the face of tumbling prices, so there are few rural jobs. With only small pockets of land to work, and long distances to travel to market on miserable roads, villagers here have a hard time making ends meet. As a result, the area has enjoyed the presence of many NGOs keen on sanitation.

Unfortunately, they have left a legacy of dependency. Villagers describe how this or that organization turned up one day, made a speech about their intentions and then proceeded to build toilets on their plots. Apart from asking the householders where to put them, no local discussions took place. Not surprisingly, people did not use the toilets properly so they stank, and children did not use them at all. In some villages this happened twice or even three times. The design of the toilets is very similar. Since in some places the water table is high, the policy has been to dig down one metre and to build up one metre. As a result, flights of steps resembling those on Mayan temples lead up to the doors of tiny cabinets erected on top of narrow plinths. No visit to the 'outhouse' could be more conspicuous, and no adjunct to peoples' homes could be more difficult to obscure by fast-growing trees or in other ways be masked or naturally absorbed into the garden plot. Where there are two or three such monuments on a property, the effect is aesthetically grotesque.

If the construction of these facilities seems ham-fisted, at least the current UNICEF-assisted ENACAL programme has been accompanied by social mobilization. According to the villagers this is the first time they have heard about the implications of toilets for their health.[38] In the small community of La Horca, the members of the Comité para Agua Potable y Saniamento (CAPS) describe the extraordinary efforts they made in order to receive the new water supply. The community had to build a road, so that the drilling rig could be brought in, and learn how to take on the management and maintenance of their solar-powered community pump. But although the water supply is the jewel in the crown, the village has also adopted 'total sanitation'. At the start of the programme, they held a community meeting to identify problems they were having with dirt and squalor: rubbish strewn around, animals on the loose, filth and ordure in the open. They then developed an action plan to deal with the situation, and elected their CAPS. 'When we knew what being dirty was about, that it was unhealthy and inferior, it had to end.' In 1999, there was a cholera outbreak in La Horca after Hurricane Mitch came through and caused much devastation, and the memories are vivid.

When asked if they see any difference in their children's health, villagers invariably say yes. This is the case not only in Nicaraguan villages, but almost everywhere where 'improved' sanitation has been adopted. In the Madagascan countryside, an extended farming family living on the outskirts of a small rural town insists that outbreaks of diarrhoea among their children have ceased. These used to be common in the rainy season, but now that faecal matter is not left lying around, and all families are utilizing their toilets, they are a thing of the past. Here again an extensive programme of home visiting and social mobilization was the preliminary to toilet construction. Many such programmes indicate that, although better health may not have been the original motivation for installing a toilet, the improvement is noticeable to the converts, especially when they know what to look out for and expect. The health benefits bear constant repetition. In El Porton, another Nicaraguan community, at least one CAPS member is charged with *charla* – speaking about health. She visits people's houses and checks that they are using the toilet and keeping it clean. When someone is recalcitrant about hygiene, the CAPS member may fetch the health visitor from the clinic to accompany her. And if a family is causing a nuisance that affects the whole community, 'we all go and clean her house. This shames her into compliance.'

Despite all the on-the-ground evidence in Boaca and Chantales that ENACAL's approach to water and sanitation is the way to go, the restructuring of Nicaraguan government departments threatens to undermine it. ENACAL is to be closed down, and instead another government organization, the public works agency FISE, is scheduled to take over responsibility. But up to now, FISE has no record of or experience with social mobilization. If the efforts of Nicaragua's international donors to articulate a common policy with the key government departments concerned with public health come to fruition, then maybe FISE will re-employ the experienced staff from ENACAL and a motivation-based, systematic, nationwide rural sanitation action can begin. If not, progress will turn rapidly into retreat. Like so many promising development initiatives, the effort to shift gears from local success to countrywide strategy is fraught with pitfalls. At one end of the spectrum, customers have to be brought on board. At the other, the conversion of the powers that be into agreeing and backing a strategy that could enable this to happen can be bureaucratically tricky and extremely time-consuming.

One country where the powers that be have made a major commitment to sanitation is Bangladesh. On the face of it, this is an unlikely candidate for 'total sanitation', especially by 2010, the target date set. One of the poorest countries in Asia, more than half of Bangladesh's people barely manage to survive on a day-to-day basis, let alone have money to spend on household improvements. Although public health engineers in rural and urban departments have done their best, governmental resources and capacity are thin. Local NGOs have also performed strongly, however, and the country has been given extensive support from external donors ever since its birth in 1971. The extraordinary inventiveness and creativity that Bangladeshis use to survive can be a powerful instrument when deployed to communal advantage.

Bangladesh is a country shaped by water. Its patchwork of land is threaded by thousands of streams and rivers, tributaries of the Ganges and Brahmaputra rivers, which every year descend from the Himalayas swollen with snowmelt, washing silt and water over the adjoining banks and into rice paddies on their way to the Bay of Bengal. This gives the country an extraordinary fertility, which allows it to support by the thinnest of margins one of the highest densities of rural population anywhere in the world: nearly 1000 people per square kilometre.[39] In this crowded landscape, a third of which is routinely flooded in the rainy season, hamlets are perched on earthen plinths, with causeways carrying roads and paths above the surrounding fields. Daily life is conducted in intimate contact with water even in the drier times of year. People bathe in ponds and streams, children swim and dive in them, religious shrines nestle beside them, fishing boats lap their banks. But these same waters are depositories for garbage and dirt. Every day, thousands of metric tonnes of shit end up on public lands or washed into these waterways. Even if 75 per cent of people use groundwater from handpump tube-wells for drinking, still their constant mingling with surface water and wastes, especially during the seasonal floods, constitute a public health problem of immense proportions. Excreta is complicit in most sickness in Bangladesh, including cholera, typhoid and all the parasitic infections, whose toll is far higher here than in other Asian countries. Around 115,000 children under five die every year of diarrhoeal diseases.[40]

Although public support for clean drinking water supplies has an even longer history in Bangladesh than does support for sanitation, nonetheless the first comprehensive effort by the Department of Public Health Engineering (DPHE) to promote sanitary latrines dates back over 30 years.[41] In 1978, with backing from UNICEF, the DPHE set up the first 100 village sanitation centres. Masons were trained to fabricate toilet components modelled on a pour-flush design originally imported from Thailand in the 1960s. As was common with early on-site models, the cost of the toilet with its concrete pit-lining rings was relatively high, so a subsidy of two-thirds of the price was offered to customers. Although take-up was promising, it was minute in proportion to the target population. But by the late 1980s, shops selling slabs, pans and concrete rings had begun to appear in the bazaars of many large towns – a sure indication that there was consumer demand among those with enough money. By 1994, the DPHE had established nearly 1000 production centres and over 3000 private producers had also set up in business. The question was how to extend consumer desire and satisfaction much more widely. This seemed to have less to do with latent demand in such a crowded environment than with the high cost of what was still an elite product.

Around this time, Cole Dodge, UNICEF's country representative, became keen to do something about the deplorable state of sanitation in Bangladesh. Despite some DPHE misgivings, it was proposed that the concrete rings to line toilet pits be dropped as a mandatory part of the design. Woven bamboo and other natural materials were thought to be adequate in most Bangladeshi soils to prop up the walls of the pit and prevent it collapsing without risk of contamination of the surrounding groundwater, thereby greatly reducing the costs. Then Dodge managed to procure political commitment to sanitation from the highest political level. The Prime Minister agreed to launch an annual Sanitation Week in which society would be mobilized around toilet construction and hygienic living. The Bangladesh NGO Forum for Drinking Water and Sanitation, a network with 10 years of experience, a membership of 560 local organizations and a well-developed infrastructure, was extremely active. The Sanitation Week was conducted on the pattern of typical Bangladeshi events, with rallies, marches, slogans, placards and songs. In the most active districts, NGOs, DPHE personnel, local officials and volunteer teams went house-by-house, holding courtyard meetings and promoting a list of key sanitation messages. Coupled with expanded production in all sales outlets, the new drive raised coverage of sanitary toilets – simple pit and pour-flush – in rural areas from 16 per cent in 1990 to over 30 per cent by 1997.[42]

Despite this progress, however, by the turn of the century there were many parts of the country with coverage levels much lower than this – often no higher than 5–7 per cent. Some toilets may have been destroyed in floods, or become full and been discarded: to establish the toilet habit definitively takes time. The volume of faeces deposited in the open landscape and ending up in waterways was still very high and as hazardous as ever, despite social mobilization methods to promote toilet use having been in use for a decade. Promotion and marketing were seen as critical, but no-one had come up with a perfect software 'fix', although many NGOs – CARE for example – experimented with health education packages and participatory approaches, and some succeeded in raising coverage dramatically in their areas of operation.

In the early 2000s, another breakthrough occurred. A local NGO, the Village Education Resources Centre (VERC), with support from WaterAid, decided to abandon exhortations to build and use toilets, and substituted the idea of 'freedom from open defecation'.[43] Communities were invited to analyse their lavatorial habits by mapping their 'defecation zones', calculating their output of excreta and the environmental health threat it posed, and then resolving to take collective action. The appeal was to self-respect and community ownership of the problem; action included naming and shaming promiscuous defecators by planting little flags with their names on in their shit.[44] The better-off members of the community, who have traditionally subsidized and managed local water-points, were now invited to do the same for sanitation, helping the poorest members by paying for their toilets. Within months, 400 villages committed themselves to abandon open defecation and started to construct toilets to confine faecal matter. This they did entirely without subsidies, a point heavily emphasized by community-led total sanitation (CLTS) enthusiasts.[45]

The Water and Sanitation Program (WSP) were deeply impressed by CLTS and began to treat it as *the* recipe for sanitation spread. During a field workshop in 2002, they brought in senior government policymakers and duly impressed them with this new 'communal disgust' or 'clean village' approach. The following year, Bangladesh declared a target of reaching universal sanitation by 2010 and adopted the 'freedom from open defecation' idea as a central component of the strategy. The next question was how to scale up a participatory approach used successfully by an NGO in a relatively small and self-contained area, and develop it as a mainstream government-backed effort. The government started by allocating resources from local development budgets for sanitation promotion, and offering communities cash rewards for reaching the 'total sanitation' goal. On the basis that nearly three-quarters of Bangladesh's rural families could not afford to buy and install a pit toilet, the government insisted that subsidies be made available – despite the conviction of hard-line NGO proponents that subsidies had created dependency in the past and that financial incentives would be the death of the participatory process.[46] But it was also agreed that the management of any subsidies should be left to local councils and sanitation task forces. During 2004, the WSP, WaterAid and the Dhaka Ahsania Mission undertook a programme of training for local councils and government bodies to activate task forces at sub-district, union and village levels.

The early results of the new approach were highly encouraging, but not extraordinary viewed on a national scale. By early 2006, around 5000 villages and 19 sub-districts had been officially declared free of open defecation.[47] Over 90 per cent of the costs were contributed by local people from their own pockets, with the government subsidy and NGOs making up the difference.[48] Subsidies had not poisoned the project as their opponents had predicted; on the contrary, they had turned out to be a useful external resource for the eradication of open defecation. More village shopkeepers and builders had entered the sanitation business, and with support from a number of NGOs, a burst in toilet construction was underway in almost every rural district. Local government officials and councils had proved capable of working with villages in a community-based approach when well-trained and receiving good support – a useful indicator that full scaling-up was potentially practicable. NGO micro-financiers such as the Grameen Bank also played an important role, helping to mobilize savings and provide small loans to villagers to finance home improvements. The health benefits were immediately conspicuous. Where villages had become 'open defecation free', they had managed to reduce from 38 per cent to 7 per cent the number of households where there had been a recent bout of diarrhoea.[49]

The Bangladesh experience with CLTS is instructive in several ways. First, it shows how important it is to gain full government backing – not only rhetorically, but in terms of financial and human resources, and to build their capacity to take forward programme implementation in a spirit of partnership. It also shows that focusing villagers on the hazardous filth in their community and their own potential role in ending this collective disgrace can be more effective in social terms than slogans exhorting the use of toilets. However, too much should not be read into the magic of this method, which is now being exported by international donors to countries such as Laos and Bolivia as the latest 'paradigm shift' in sanitation. The reasons for success in the Bangladeshi environment may be as much to do with the crowded landscape, people's sense of personal need, traditional patronage relationships, over 20 years' repetition of sanitation and health messages, and a lively history of mini-entrepreneurship – all of which are factors external to the programme design – as to the discovery of a sanitary golden key. The government is not convinced that CLTS is able to solve the problem of inequity: however cheap a low-cost toilet becomes, it remains unaffordable for the poorest households in a country where many families can barely afford to eat. Expecting the better-off in each village to pay for poorer households to have toilets so as to dispose of all faecal threat to themselves is sensibly regarded by the authorities as an unreliable strategy.

There are many reasons for Bangladesh's sanitary progress. Because it has such a high diarrhoeal caseload, decades of effort have been put into sanitation in Bangladesh, particularly by NGOs and the DPHE. The government – whatever its other deficiencies – has been open to external ideas and singularly enlightened about faecal issues, given its modest means. And there are other special local assets besides that of official and political backing. Micro-finance was invented here and the Grameen Bank first started giving loans to the very poor for housing improvements, including domestic handpumps and toilets, well over two decades ago – a situation which does not prevail in other countries. All the post-Water Decade activity – social marketing, training of local sanitary entrepreneurs, the annual Sanitation Weeks, a strong emphasis on knowledge spread – helped pave the way for a new sanitary lift-off. Time, familiarity and lessons learned during earlier programmes are of the essence in any context where a major behavioural change is being sought. Progress towards 'sanitation for all' in Bangladesh could easily languish once again if political and popular enthusiasm drops off. What is remarkable is how far Bangladesh has managed to go with sanitation in a generation – a far shorter period than the average sanitary revolution in Europe and North America. In 1985, the popular wisdom was 'marry your daughter to a man with a handpump tube-well'. Two decades later, it had become 'marry your daughter to a man with a sanitary toilet'.

There is no 'perfect' approach towards rural sanitation, any more than there is in any other development context. Indeed, in some contexts it may be best to leave well alone, or simply add scientific knowledge about shit and pathogen avoidance to the local school curriculum (see Chapter 5). If pit toilets are installed badly or not kept clean, they may actually help to spread infection.[50] Where there is definite user-based need, there is no one programmatic solution. Approaches which have worked well in one place cannot be transplanted to another without sensitive and creative adaptation – although there are some useful principles to be learned from the many experiences which now exist in different settings. What is noticeable is, however, that in every case there is a strong existing social incentive: lack of 'space', lack of vegetation and therefore of modesty, crowdedness, female insecurity, and a strong sense of disgust that can be enhanced by knowledge and ideas penetrating from elsewhere.

In every setting, to be successful, an approach will have to take account of local considerations: beliefs, income levels, costs, political and popular attitudes, and the availability of official and external support. Comparisons between Bangladesh's recent 'total sanitation' success with programmes in other countries are invidious, except when made with West Bengal, where there are genuine social and environmental congruities: in both, rural demand for sanitation is potentially high because of specific, and almost identical, circumstances. Once Bengalis have been confronted with the risks they face from open defecation, they are primed to abandon it, if they can conveniently do so within their means, or with some kind of assistance if they are extremely poor. Better health may not typically be the primary motivation, but the potential health benefits can exert a powerful influence where they are convincing and real – in Bengal and elsewhere.

Finally, the question of the consumer item itself – its nature, practicality, costs and appeal – should not be underestimated as a critical ingredient. Much opprobrium has been poured on 'supply-led approaches'. But if the right kind of water-flush toilet and sewerage system had not come along in the 19th century, the people of the industrialized world would still be using 'dry conservancy'. There is not much point in cultivating demand for something unless supply is also taken care of. When it became clear to the sanitary pioneers of the late 20th century that the pit toilet would have to be the genus of toilet used by the vast majority of the world's inhabitants, they put a great deal of energy into its improvement. The modern story of the outhouse, dunny, earth closet, VIP, twin-pit and pour-flush is where we turn next.

4

Pit Stops: The Expanding
Technological Menu

Previous page: A woman in Zambia pictured rinsing her hands from a water container outside her VIP toilet. Construction costs have been kept down by use of local bricks, the simplest of vent-pipes, and by avoiding the need for a door. Overhanging thick thatch keeps the entrance dark and the interior cool.

Source: Jon Spaull, WaterAid

The simple pit, with planks on top, over which to squat while evacuating the contents of the larger intestine, is the oldest and most widely used toilet in the world. Sewerage in ancient Mesopotamia, Harappa, Crete and Rome is more celebrated because it demonstrates that early civilizations had a sophisticated grasp of hydraulics and engineering. Systems with chamber pots and buckets required servants or sweepers to empty them. But the outhouse or 'necessary room' was self-contained: if the pit was deep and wide enough, the liquid content leached away into the surrounding soil and the remainder rarely needed emptying. Near the river bank or beach, the pit could be dispensed with altogether. Planks or trees were stretched across ditches and crude platforms constructed on stilts above streams to create what are known as 'hanging latrines'. In places where water did not flow conveniently past, and among people who were not living in monasteries, forts, castles or palaces, with servants to carry away their 'close stools', but in simple town houses or cottages, a pit in the ground, capped with a shelter, was the only sanitary alternative to the great outdoors. Hence the pre-industrialization preoccupation with cesspools, 'middens', dunnies and jakes, or however and in whatever language the house of easement was colloquially known.

So scant is colonial literature on lavatorial life that some guesswork is required concerning the earliest attempts to improve 'on-site' toilets in the developing world. Sanitation was undoubtedly a concern of 19th-century colonial administrations, especially in the Indian sub-continent, where cholera had first been identified and from where it not only spread around the globe, but also caused havoc in its country of origin. But as with their municipal counterparts back home, drainage, sewerage, and the control of public nuisances in rapidly expanding cities such as Bombay, Madras and Calcutta were the colonial engineers' predominant interest.[1] In its passage to imperial lands the flush toilet took up residence in colonial mansions and probably in those of local grandees. In most of Asia, local systems were 'off-site': collection involved the employment of carters or sweepers to take away ordure by the bucketful. Of toiletry in pre-European African civilizations, little is known. However, a colonial visitor to the Asante capital, Kumasi, in what is now Ghana, noted in 1817 that every house had its 'clocae, besides the common ones for the lower orders without the town'.[2] Sanitation and cleanliness in Kumasi at the time were exemplary, and still impressing visitors in 1874. Elsewhere in the continent, some water-borne or flush systems were built in elite quarters of towns by Europeans, and in missions, colleges and offices. But pits and buckets – buckets were regarded as an improvement over pits when introduced into Maseru in southern Africa in the 1930s[3] –

seem to have been the only dedicated facilities provided by colonial builders for Africa's indigenous inhabitants.

In 1897, a British engineer invented the septic tank, which upgraded on-site possibilities and became the aristocrat of unsewered sanitation; with a tank and pull-handle or pull-chain flush, a toilet connected to a septic tank is indistinguishable from a sewered WC. But in many environments, there is a dearth of water for tank-and-pull mechanisms, not to mention a lack of resources for the necessary standards of housing construction and internal plumbing. Some simpler version of the water-flushed toilet was required; hence the development of the pour-flush pit toilet of Asia. The bowl or pan of this toilet is fixed over a pit and water-flushed by hand using a bucket, without any intervening mechanism, chain or handle. The flush does not have to remove cleansing materials from the pan because most people in Asia clean themselves with water. The pour-flush toilet can be seen either as a lower-class version of the WC, or as a superior kind of latrine (it is usually demeaningly known as a pour-flush latrine). In India, according to A. K. Roy, an Indian sanitation guru of the 1970s, the pour-flush water-seal latrine was first developed in the mid-1940s at the All-India Institute of Hygiene and Public Health (AIIHPH) in Calcutta (Figure 4.1).[4]

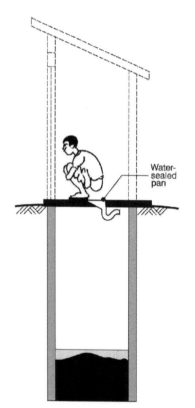

Water-sealed pan

Figure 4.1 *The simple single-pit pour-flush pit toilet*

Source: WEDC, © Rod Shaw

However, a country with a superior claim to the invention of the pour-flush is Thailand. Certainly, there was more success in Thailand – or Siam as it was then known – in popularizing pit toilets as compared with India during the British period; Siam was never colonized, which may have something to do with it. Public health activity began, here as elsewhere, in the city: Bangkok's first sanitation law was passed in 1897, to regularize garbage collection and construct public lavatories. In 1924, the Governor of Sukhothai Province, a gentleman named Sawadi Mahagayi, invented an improved latrine for household use: the 'goose-neck' water-seal pour-flush toilet.[5]

Mahagayi fitted a short pipe on the outlet of a toilet bowl, doubling back upwards to hold the water in a U-cup or 'trap'. Inverted, the component looked like the head of a long-necked bird – hence its name. Sanitation was taken up vigorously in Thailand, with the support of no less a personage than the father of a future king, Prince Hahidal of Songkhla, who exhorted the people to adopt sanitation 'to control the exits and entrances from and to the human body'.[6] Decades of effort led to 98 per cent coverage in Thailand, and their invention of a simplified WC also became the basis of sanitation spread throughout much of rural and semi-urban Asia, including Bangladesh, India, Myanmar and Indonesia. Where porcelain is not affordable, the bowl is fabricated separately in a hard material, dried, sanded and polished to make the surface slippery; its gooseneck is then attached, and the whole set into a slab. The water-seal vastly improved the lowly status of what had previously been a simple pit, closing off smells and distaste-fulness.

The pour-flush gooseneck accommodated cultural ideas and toilet habits and kept the costs and complications of sanitation spread relatively low. The only other external difference from a typical European WC was that the pan was usually at floor level: toilet users in Asia tend to squat rather than sit on a pedestal, which is an unfamiliar experience to people who normally sit on the floor. Many peoples not only in Asian cultures but also in Africa are also averse to sitting on a toilet seat because of the indirect contact with other naked bottoms.

A different kind of toilet was introduced in another part of Asia. In the early 1950s, Dr Nguyen Dang Duc of Vietnam designed a double-vault composting (DVC) latrine (Figure 4.2); from 1956 onwards this was heavily promoted in the northern part of the country.[7] The use of the DVC was part of a government effort to improve environmental sanitation in the deeply impoverished country-side by means of what was then the standard approach to public health in this part of the world: the mass campaign. Slogans such as 'clean house – fertile fields', and 'building three sanitary works: water wells, bathrooms and toilets' were used.[8] Here, as in China, the use of human excreta for manure was routine, so the toilet had a dual purpose: sanitary confinement of pathogens until they were well and truly dead and collection of excreta for agricultural use. The twin vaults or chambers, which were built above the ground on a solid concrete base, were used solely for faeces, and the whole point was to use the vaults alternately so as to allow the content of each to be closed off for several months before spreading the fertilizer on the fields. Thus transformed to compost, the dry product had the same nutrient properties as faeces without the public health hazard or unpleasantness.

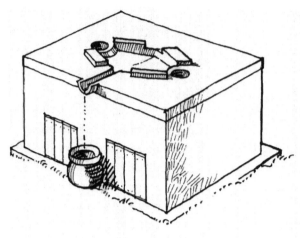

Figure 4.2 *The double-vault composting toilet, Vietnam*

Two processing chambers are provided with a squatting slab for urine diversion, a pot for collecting urine and doors for the two openings for removal of dehydrated material. The drop hole not in use should be closed with a stone and sealed with mud or mortar.

Source: Mayling Simpson-Hébert and Uno Winblad (eds) (2004) *Ecological Sanitation*, revised edition, Stockholm Environmental Institute, p23

However, war intervened and disrupted the sanitation campaigns of the 1950s and 1960s, as it did civilian administration and rural life generally. When double-vault composters were revisited after the Vietnamese conflict ended, it was found that farmers were not managing them properly. They opened up the vaults to take out the contents when they needed them without waiting long enough for pathogen destruction. The knowledge and skills needed for correct DVC management had not been passed on. An effort was made to introduce the pour-flush – a superior toilet from a hygienic and aesthetic point of view – but farmers frequently broke off the water-seal or otherwise damaged the slab and pan to gain access to the manure, and did whatever was necessary to convert their pour-flushes to ordure-collecting dry toilets.[9] The double-vault composter was shortly to enjoy a renaissance, as we shall see.

Meanwhile in Africa, efforts existed to promote sanitation in rural areas during colonial times, especially village cleanliness. 'The sanitary inspector' of rural Nigeria was a familiar and redoubtable figure who imposed fines on dirty compounds for helping insects breed disease. But there was a tendency not to take him very seriously. Ken Saro-Wiwa, the Nigerian writer, wrote a short story called 'The Inspector Calls', in which the sanitary inspector visiting the village of Dukana on his motor-bicycle manages not to see the dollops of faeces of children, dogs and goats littering the footpaths, and instead is royally entertained by the Chief and his entourage. He leaves the village burdened with gifts, taking away in the dust clouds billowing up behind his departing bike 'all those ill winds which, had they remained behind, would surely have plagued Dukana'.[10] The sanitary inspector represented the intrusive agent of colonial, or modernizing, forces. The idea that white colonials could teach Africans how to be 'clean' was frequently regarded as offensive.[11] Suggestions concerning the construction of toilets for confining human wastes were still being greeted with derision in parts of rural Nigeria in the late 1980s.

Positive pit toilet news from southern Africa starts in the 1970s. The more pervasive character of European settlement in the southern part of the continent exposed local people in countries such as South Africa, Swaziland, Botswana and Lesotho to the idea of using a toilet earlier, and to sitting on it, in a way that was less common in most other parts of sub-Saharan Africa. Among southern African populations, sanitation issues arose from the livelihood and demographic upheaval represented by the colonial precursor of 'development'. Africans were coerced, taxed or otherwise manipulated into working for white employers in mines and on plantations or farming estates, but also as the lowest-level performers of all 'modern' productive or domestic activity. The millions of workers in all parts of the region who left their mud and wattle compounds for most of the year to live in barrack-type hostels provided by South African mining companies; who occupied service quarters behind city homes in Johannesburg, Salisbury, Blantyre, Laurenço Marquez, Gaberone or the copperbelt towns of Zambia; or who lived in congested rural slums on the edge of highly managed tea or coffee plantations, farms or forestry plots became familiar with toilets, even if many were very crude. In the less European-controlled and more dispersed rural way of life in which most sub-Saharan African people lived, they would not normally have encountered this curious item, nor experienced the conditions of living to which its development in industrialized settings had responded.

It is not surprising to find, therefore, that the earliest efforts to improve pit toilets come from this part of Africa, specifically from Zimbabwe, then still Rhodesia. Unlike the early origins of the improved 'wet' latrine, the pour-flush of Asia, the story of the improved 'dry' latrine, the VIP of Africa, is much better known. The key R&D work was undertaken by Peter Morgan, a pioneer who is definitely to be ranked as one of the heroes of the modern low-cost sanitation story – a late 20th-century equivalent to the Reverend Henry Moule in Victorian Britain, who patented his earth closet just over 100 years earlier (see Chapter 1).

The toilet invented by Morgan in 1973 was originally known as a Blair, after the institute where it was developed, itself named after Dr Dyson Blair, a former secretary of health in Southern Rhodesia and a keen advocate of the health benefits of low-cost sanitation.[12] Ever since the 1940s, there had been active work by the Environmental Health Department and local 'health assistants' to promote hygiene in rural areas of the country. Brick-built pit toilets were already becoming quite common in the 1960s, before Morgan began work on a new and better version.[13] In 1975, after two years of tests, the Blair was taken up by the Ministry of Health. This was a period of conflict in Rhodesia, as African liberation forces attempted to wrest political control from the white minority regime. Although the war hindered its rapid promotion, the Blair was used in residential quarters on farms and estates and in the congested conditions of 'protected villages' which many rural people were forced to inhabit.[14] These concentrations of people faced obvious public health threats and cholera outbreaks were not unknown. Once independence arrived in 1980, along with much-needed donor support, the new-style toilet became established as a cornerstone of rural Zimbabwean public health. At the dawn of the Water and Sanitation Decade in 1981, the Blair slipped naturally into place as one of the front-running promotional items of sanitary ware.

The key feature of the improved dry pit toilet was that it used natural processes of wind and light to reduce the presence of smells and flies. Freedom from odour made the toilet nicer to use than the standard pit latrine, and freedom from flies boosted its disease-control qualities. The cleansing materials could be leaves, corncobs or other natural and degradable materials. The key improvement was a vent-pipe rising from the underground pit (Figure 4.3). Wind passing across the top of the pipe had the effect of creating an up-draught which sucked the smelly air up and out. The hole or pedestal used for deposits needed to be left open to create an inflow of air and the cabin needed to admit air to funnel the draught. The system of fly-control was similarly prosaic and similarly efficient. Passing flies were attracted to the evacuating smell, but a wire mesh screen over the top of the vent-pipe prevented their entry. If the cabin was kept dark, any flies that found their way inside would be attracted towards the light at the top of the vent – where they would find their exit blocked, and eventually succumb to exhaustion. Morgan's tests at the Blair Institute found that, over 78 days, nearly 14,000 flies were caught in an unvented pit toilet, whereas in an identical toilet with a vent-pipe the count was only 146.[15] The toilet's ventilation feature was defining, which is why outside Zimbabwe, the Blair became known as the 'ventilated improved pit' or VIP.

In the mid-1970s, while Morgan was tinkering with his Blair, a handful of other engineer-humanitarians were becoming interested in low-cost sanitation. One of these, John Kalbermatten, Senior Water Supply Advisor at the World Bank, set out to change attitudes towards public health at the international level. Kalbermatten was one of the first experts to emphasize to policymakers in influential places that investments in sewerage were never going to reach the world's

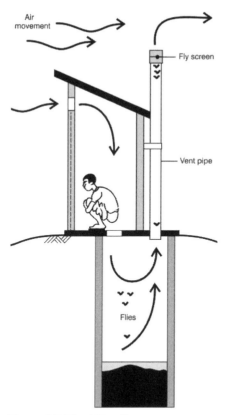

Air movement

Fly screen

Vent pipe

Flies

Figure 4.3 *The VIP toilet*

Source: WEDC, © Rod Shaw

poorest members.[16] In 1972, Robert McNamara, President of the World Bank, had declared that the Bank's programmes should address poverty directly. Kalbermatten argued that unless alternatives to sewerage were identified, this would not be possible in the context of sanitation.

In 1976, he won agreement to establish a multidisciplinary research project within the Bank to look into the kinds of technologies which could be used for water supplies and sanitation in the poorer parts of the world.[17] The project explored whatever was out there in terms of appropriate technologies for low-cost, on-site sanitation, including the work of Peter Morgan at the Blair Institute and Indian enthusiasts Bindeshwar Pathak and Ishwarbhai Patel, keen protagonists for replacing bucket and sweeper systems with pits and flushes. The technologies considered by the Bank project were the VIP, the pour-flush, the composting toilet, small-bore sewerage systems and septic tanks. This helped prepare the ground for the Water and Sanitation Decade; indeed John Kalbermatten was critical in lobbying for the Decade, anticipating the momentum and funds it would generate as the springboard for the new sanitary mission.

The next step was to obtain practical support for programmes based on the appropriate technologies identified. UNDP, the lead UN agency for the Decade, was invited to join in a partnership with the World Bank, out of which came the international Water and Sanitation Program (WSP). This was the first serious indication by large-scale players in international development that – should these basic technologies prove viable – it might be possible to develop a new sectoral approach for low-income communities on a non-industrial, non-water-borne sewerage basis. In 1978, the first global project of the WSP was set in motion: the creation of a Technical Advisory Group (TAG). The TAG was to study, pilot, and lay the technical, institutional and financial groundwork for on-site sanitation systems suitable for use in poor urban and rural areas. If there was really going to be a new sanitary revolution, the technologies on which it would be based would require the credentials necessary to bring on board donors, private investors, teaching institutes and public health engineering bodies. Hopefully, once they had been given the imprimatur of a World Bank-backed operation and high-powered sanitary experts, large-scale investors would pick up these currently small-scale and idiosyncratic new approaches and enable them 'to go to scale'.

The vision of the TAG was ambitious; nearly 30 years later, it is still ambitious. In spite of everything that has since been achieved, the full absorption of low-cost technologies into the engineering and corporate sanitary mainstream has still to occur. How and why this purpose has been blown off course will be examined later; for the moment, it is enough to point out that – like their 19th-century forebears – the sanitary pioneers of recent decades have faced extraordinary frustrations in trying to gain for their mission the seriousness and mass uptake it deserves.

One of the countries to attract the early attention of the TAG was the tiny kingdom of Lesotho, enclosed within eastern South Africa. Driven by population pressure to cultivate higher and higher in the mountains, the Basotho people made a hard living out of agriculture, and 40 per cent of the men migrated to work in South African mines and industries.[18] The British government, from which Lesotho gained independence in 1966, was anxious to help its poverty-stricken ex-colony. In 1975 it funded an evaluation of Lesotho's water programme by Richard Feachem of the London School of Hygiene and Tropical Medicine. Feachem, another influential figure, proposed across-the-board improvements in water, sanitation and hygiene. His call for an integrated programme was echoed by the TAG, which between 1978 and 1983 sent a series of missions to Lesotho. Keen to promote the Blair, newly dubbed the VIP, the TAG saw Lesotho as a testing ground for its attempt to convert the sanitary establishment to its on-site creed. The United States Agency for International Development (USAID) was also active in Lesotho, and the TAG's influence led to a concerted agreement that sewerage was economically out of the question even for urban schemes, and that the VIP would be introduced instead. This led to the most renowned early sanitary success of the Water Decade.

By 1983, when the rural sanitation pilot project in Lesotho started up with assistance from UNICEF and UNDP, the VIP was a tried and tested toilet. In

Zimbabwe, tens of thousands of Blairs had been constructed and Ministry of Health staff had become thoroughly conversant with the technology. The earliest versions were rectangular, made of ferro-cement and had wooden doors; but eventually a spiral shape that did not require a door was developed instead. As the Blair became more widely adopted, there were many experiments with natural materials such as grass, timber and plastered mud. But eventually the Ministry of Health chose a brick-built model. People in rural areas were used to making bricks, they were suitable for every terrain, and it was thought that the longer-lasting structure would help instil the change in sanitary behaviour necessary to make public health benefits sustainable. The downside of this decision was that, since each Blair built according to Ministry specifications was relatively expensive, most people could not afford one. Thus the only way to spread the Blair nationwide was to provide a substantial subsidy. This was the 'supply-driven' sanitation model for the post-independence programme, under which construction of VIPs went ahead vigorously, peaking in 1987, when 50,000 altogether were built.[19]

Not only Zimbabwe but Botswana too had extensive experience with the VIP by the early 1980s. The Botswanan Ministry of Local Government made their own adaptations and conferred their own designation on what they believed was a superior version: the BOTVIP.[20] One of the main points of consumer resistance to dug toilets is the fear that the pit will collapse, landing the user up to the neck in mire. This prospect is particularly alarming to children. If the pit is in soft or crumbly soil and has to be lined with bricks or other material to withstand the weight of the user and structure above, costs are inevitably increased. On the other hand, if the ground is very hard and rocky, it is extremely difficult to dig down to any reasonable depth (three metres was the common recommendation). One of the key design changes in the BOTVIP was to offset the pit, so that only a small part of it was underneath the superstructure, and a cover was put over the exposed part of the pit top so that it could be emptied. The introduction of the offset pit was an important development in VIP technology and was subsequently exported to Nigeria and other African destinations.

In Lesotho, a number of VIP variations were considered, relating to soil conditions and to the constant effort to reduce unit costs. Unlike in Zimbabwe, where costs were not such an issue in the early-1980s, the cash-strapped government of Lesotho insisted from the outset that customers for VIPs would have to pay for toilet installation. Indeed, the government can be seen in retrospect to have been very avant-garde in expecting all construction to be handled by the private sector. The authorities undertook the training of local masons as prospective toilet-builders, who were then hired out by householders at agreed rates. Thus the idea from the first was to integrate toilet construction into the local economy. The programme worked with community leaders to make sure local management was well-oiled, with health centre personnel running hygiene education campaigns.

The approach worked extremely well during the three years of the pilot in Mohale's Hoek District in the south. The lack of local jobs was an incentive to young entrepreneurs to enter the toilet construction business. Armed with a certificate, and with health workers promoting sales, graduates were expected to drum up their own trade. And this they managed. Instead of the 400 VIPs targeted, by 1986, 600 had been built, many fully paid for by householders. In keeping with the latest thinking, women were very much involved in the programme, not only as health promoters, but also as builders: one builder in four was female.[21] Women tended to charge customers much less, especially if they were poor, on the basis that toilets were a public service. Partly due to proximity to South Africa and familiarity with toilet notions there, and due to the well-crafted approach, sanitation in Mohale's Hoek took off. In 1987 the programme went national. To begin with, the national programme was very dependent on external donors, but the government was sufficiently committed to incorporate the strategy into the 1989–1990 National Plan and assume a greater share of the costs. However, in 1990, when international enthusiasm surrounding this sanitation story was at its height, the notion of self-supporting, consumer-driven toilet provision in rural Lesotho was still quite distant: as is often the case, an incipient success is often written up before it occurs. The donors' expectations that, within a few years, the kingdom would be effectively fully VIPed turned out to have been over-optimistic, but the government has remained committed, providing strong budgetary support.[22]

The VIP did face opposition from some quarters, particularly those which regarded any form of sanitation technology associated with the words 'pit' and 'latrine' as inferior. There were frequent claims that the pit contents would contaminate surrounding groundwater. However, unless the cubicle is used as a shower or as a wastewater disposal unit – a wrong use of a pit toilet – the amount of seepage is small, and distancing the toilet from local water sources by several metres turned out to be sufficient. There were also objections that the space required eliminated the use of VIPs for crowded towns and cities. But the much greater difficulty, one that was to prove a constant bugbear of the VIP every- where it went, was the question of cost.

Unlike a WC installed in a home, the VIP normally has its own freestanding cabin. This was partly to fit in with African living patterns, and partly because the top of the vent must be exposed to wind. In traditional Africa, housing was – is – of mud and thatch, with a group of buildings comprising a compound. Family life is based on the compound, not the house; and the compound can contain many sorts of buildings – stores and kitchens, for example. Space is usually not a problem; although this is beginning to change in densely populated areas. Until recent decades, the idea of cash expenditure on rural housing – for example on a tin roof or breeze-block walls – was relatively strange, reserved for the rich and status-conscious. In most households, dwellings were repaired with natural materials such as mud and thatch once a year.

So expecting people to pay money – that valuable and scarce commodity – on building a solid 'toilet house' was unrealistic for the average household; only a chief, or a professional or business person, might do such a thing. In Lesotho, with its mountains and freezing cold temperatures, expenditure on thicker, stouter, burnt-brick walls was less surprising. But however much costs were kept low by using local materials for the cabin, a basic minimum of expenditure was needed: on cement, pipes, mesh or gauze, and – where the soil was unstable – pit lining. If bricks were to be used for the cabin and could not be made by the householder but had to be bought, these were also costly. But if their number was cut too far, the cabin would be hot and claustrophobic, testing the deodorizing qualities of the 'ventilated' in VIP to their limit. And besides, where people did invest money in a 'toilet house', prestige was often the motivation. Since only the upper works could be seen, the status-conscious owner might not want an inferior kind of structure but the smartest and most solid. These variations in consumer style and motivation needed to be taken into account. There were cases – in Togo, for example – where VIPs were allocated exclusively for use by distinguished visitors.[23]

In Lesotho, a market survey by USAID suggested that around 45 per cent of the rural households could afford a VIP; a further 30 per cent would need credit, and 25 per cent would need a partial or full subsidy.[24] Local credit unions were prompted to extend loans for toilet construction, but this did not help the poorer households. And in Lesotho, a real demand had developed for VIPs, which was not the case in most of the rest of Africa, especially in countries farther away from the south. Even in Zimbabwe, where originally up to two-thirds of the cost was met by a subsidy, the remaining amount was still unaffordable for many people. And subsidies threw up another problem: they were often captured by better-off households or those with 'connections', meaning that there were fewer resources to help poorer households.

In the late 1980s, the Mvuramanzi Trust, a Zimbabwean NGO, reduced the VIP subsidy to three bags of cement and a fly-screen, equivalent to US$14; the

Zimbabwean government – anyway beginning to backtrack from heavily subsidized programming in the face of budgetary squeeze – followed suit as it became increasingly clear that the role of subsidies in toilet promotion could actually be inhibitive. As cheaper VIP models were promoted, subsidies dropped still further – to around US$10 in the 1990s. But despite the drop in VIP costs and prices, with the exception of Lesotho, there is nowhere in the world where VIPs in significant numbers have been constructed by family consumers purely on the basis of market demand. So distinctive are the chimneys of VIPs in the picturesque Lesotho landscape that they are even written up in international travel guides.[25] But whatever the special factor was – the cold weather or habits learned in South Africa – it does not have that golden asset: potential for replication elsewhere.

This is not to decry the success of VIP technology, which has been more widely used in schools and other institutional contexts than by households. In Zimbabwe, the key ingredients of VIP construction in a variety of formats – single pit, double pit, offset pit, deep pit, shallow pit, spiral cabin, rectangular cabin, bricks, mud-plastered pipes, chimneys – are known throughout the local building industry, and every child leaves school understanding the lavatorial value of a vent-pipe and fly-screen.[26] However, the cost of even a relatively modest VIP has reduced the attention it receives today as the key to mass sanitation. In countries that have suffered economic downturns, such as Nigeria, sanitation experts have sought cheaper alternatives than the VIP to promote in rural areas.[27] The costs of separate household VIP installations in many urban areas may actually be more, administratively and promotionally, than those of the latest small-bore sewerage systems. Much depends on settlement density, compound size, specific costs and preferences. In one Ghanaian town of 600,000 – Kumasi, where toilets of a pit variety have been used for at least 200 years – VIPs were chosen over sewerage in the early 1990s mainly because the costs of sewerage were prohibitive by comparison.[28] By contrast, concentrated urban populations in Senegal today prefer a pour-flush water-seal toilet, a type of device not on offer in Kumasi at that time.

In Zimbabwe, the country which most closely embraced for rural populations the sanitary technology it invented (Blairs are not allowed in towns), a UNICEF-sponsored inventory undertaken in 2004 put the total of family Blairs at 422,378, serving an estimated 2.1 million people.[29] Yet however impressive, this only represents a coverage figure of 24 per cent, and this proportion, due to the problems of acute economic stress recently faced by Zimbabweans, has since declined. In Lesotho in 2004, 40 per cent of the rural population had access to 'improved sanitation', showing that here too universal coverage has not been achieved.[30] The VIP has been exported all over the developing world, but in relatively few of the destination countries – Uzbekistan and China are two exceptions – have these toilets been extensively taken up for home use. However, one organization dedicated to promotion of VIP technology over many years in southern Senegal is beginning to find that rising educational and income levels are starting to turn householders towards this solution to their excretory needs.[31] The VIP's fortunes may well see an upturn not only in parts of Africa where living standards are improving, but in parts of Asia too. In developing country settings, after all, a generation is not a long timeframe for the transfer of a technology requiring radical behaviour change.

It bears repeating that the inhibitions to sanitation spread are much more complex than those associated with the technology, especially in much of rural Africa. Built and used as it is meant to be, the VIP is a five-star toilet and well suited to water-short areas, as has been proved in southern African settings. Consumer take-up outside sub-Saharan Africa, however, has been artificially limited by the failure to expose the technology to potential promoters and customers. VIPs and other dry types of toilet have rarely been promoted in those parts of the Indian sub-continent suffering desert conditions, for example, for reasons of 'cultural inhibition' (see Chapter 6). But there are good prospects with the necessary promotion. In Afghanistan, for example, VIPs and dry composters are gradually gaining ground.[32] Despite the failure of the VIP to fulfil the TAG's vision of mass adoption and institutionalization in the African mainstream, it is important to note that the influence of Morgan–Blair thinking on later, simpler and cheaper on-site toilets all over the world has been profound.

By comparison with the low-cost device invented in Mozambique in the early 1980s, the VIP was a palace. While Zimbabwe emerged from civil war, Mozambique remained engulfed in internal conflict until 1992 and suffered huge social and economic destruction. However, even during the war, the government placed a high priority on sanitation. Townspeople were encouraged to build latrines and did so, but they were all of the old, unhygienic variety and liable to collapse. An effort was therefore made to design a safer, more hygienic toilet that would be acceptable and affordable to most families. What people in Mozambique said they wanted was not a fancy superstructure, but a solid cover for the pit. Hence the 'sanplat' was born. This was a concrete cover or sanitary platform, slightly domed to help with cleaning, with foot-rests raised to position the user and a hole shaped like a key-hole with a tightly fitting plug.[33] The circular slab was fabricated using a mould. Within a few years 11 local production centres had been established and sales in and around Maputo had taken off.[34]

On the strength of this success, a national low-cost rural sanitation programme was launched in Mozambique in 1984. Slab workshops were set up in new urban locations, and by 1987 production rose to 25,000 a year, although since the workshops were in town centres, transport remained a problem. The programme turned out to be too dependent on external funds (mostly from UNDP), however, and too technically oriented. The costs of materials for the slabs soared, and in 1988 the government introduced a subsidy. From this point on, the 'sale' of sanplats in Mozambique waxed and waned, according to whether they were subsidized and by how much. Here was a lesson to be absorbed. Sanitation's failure to take off in the country was ascribed not to the technology but to the lack of 'demand creation'. The dependence of its fortunes on subsidies and a 'supply-driven' model helped to reduce its reputation. In 2002, an estimate pointed out that, at the current level of subsidy, around US$60 million would be needed to provide slabs for the 2 million Mozambican families without toilets.[35] Money was not the only problem: it had become widely agreed that without real behavioural change, toilet use was a ten-minute marvel. If it broke, or became full, the toilet was abandoned. In Mozambique, insufficient thought had been given to the problem of pit emptying. Unless needs were satisfied on an ongoing basis, toilet demand, behavioural change and sustainability could not be collectively assured.

Whatever its mixed fate in Mozambique, the sanplat has been another successful sanitary step forward; its Swedish inventor, Björn Brandberg, who now promotes it commercially worldwide by marketing 'moulds and manuals in a box', claims that a hygienic sanplat can be produced for only US$2.[36] In the years since its arrival on the sanitary scene, between 3 million and 4 million sanplats have been installed, according to his estimation. The simplest version is a 50 or 60cm square slab, placed over a pit in a timber or mud toilet floor (Figure 4.4). The smoother and more polished the sanplat, the higher its status, and the easier to clean. Brandberg's latest model – the SaniPlast Privé – is made of coloured plastic and can easily be transported on a woman's head in the time-honoured African method of porterage. It has another refinement: its lid can be lifted off the key-hole with the foot, avoiding the contamination of hands by stray faecal particles. Although the plastic version is more expensive, it has more appeal as a consumer item – an aspect explored further in Chapter 5.

Most rural programmes in Africa use some version of the sanplat, with a little VIP thrown in if people can afford it. In the Djourbel district of Senegal,

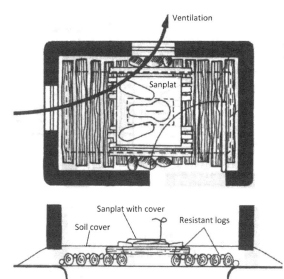

Figure 4.4 *A simple sanplat*

Source: Bjorn Brandberg (1997) *Latrine Building*, Intermediate Technology Publications, London

for example, it can be found in a large spiral enclosure made of brush and open to the sky, with a narrow vent-pipe anchored within the slab, jutting up like a telephone pole. This reduces the build-up of smells in the pit and provides some fly control, although not as efficiently as the roofed VIP. Mothers and children prefer them to the cramped brick boxes of the past, which they have now turned over to their menfolk and to visitors. Both here, in its more spacious version, and in highland Madagascar, where the sanplat toilet has to be housed against the rain and has no such frippery as a vent, such simple toilets are well received, used by all the family and usually kept clean. But the cost is almost invariably subsidized.

In many rural environments, the majority of people living at or close to subsistence level cannot be expected to install a major item of household improvement, costing more than any building they have previously erected, without some form of financial assistance. In Tawafall village in Djourbel, for example, where the women's '*Association de courage*' has organized a sanitation programme to keep the village clean, by no means all of the members yet have vented sanplats or even old-style unimproved latrines. Here, the total costs of the new-style vented sanplat toilet are around US$35, and the amount expected from the household – not including digging and transport – around US$20 (CFA10,000).[37] Even this amount is impossible to raise for many women. Maty

Fall, the *Association de courage* Treasurer, whose husband is old and infirm, whose children are very young and who scrapes a living growing ground-nuts, speaks for the members when she says, 'Everyone here is in favour of toilets. It is just a matter of having the means.'

In sanitation, the plain fact is that even the very cheapest technology, in an environment where modern technology of any kind is a rarity and livelihoods are still tied to natural resources and bare subsistence, is expensive. How to introduce into a pre-industrial economy improvements which involve manufactured items or equipment, whether in the context of food, water, sanitation or anything else, and yet avoid dependency or non-sustainability remains a puzzle. In such a setting, it is reasonable that people put other basic needs ahead of improved sanitation.

At the other extreme of the pit toilet hierarchy is the pour-flush, to which we now return. When members of the World Bank's research project into alternative technologies arrived in India in the mid-1970s, they were extremely impressed by the public facilities built by Dr Bindeshwar Pathak. Dr Pathak had founded his 'sanitation movement' – Sulabh International – in 1970, and the centre-piece of his efforts to extend toiletry in urban areas was the *shauchalaya*, a twin-pit version of the pour-flush water-seal, which he himself designed (Figure 4.5).[38] A pipe leading from the pan forked left and right, and one branch was closed off depending which pit was in use. The two pits were used alternately; in this way the contents could be rendered into harmless compost and no-one need handle the wet faeces, a key aspect as far as Pathak and other campaigners against manual scavenging were concerned (see Chapter 6). This toilet, which Pathak promulgated as a public, pay-as-you-use facility, was taken up by the TAG as *the* toilet for Asia, in the same way that they took up and promoted the VIP for Africa.

At this point in the Indian sanitation story, there had been no effort to promote low-cost sanitation in rural India except by NGOs. Their experience showed that any such enterprise needed to build on demonstrated demand; the provision of a free toilet without any attempt to motivate the recipients virtually guaranteed its lack of use.[39] At the time, efforts were being made by water and sanitation donors, who set up a special Indian TAG, to persuade the Indian government to make a serious effort to live up to its Water Decade commitment to raise rural sanitation coverage from almost nothing to 25 per cent. However, the Indian TAG – in which senior figures from Indian research institutes were involved – became carried away by technical issues and, ignoring the NGO experience, failed to embrace aspects such as community participation and consumer demand.[40] Most of their efforts were put into technical refinement of the toilet itself. The idea of standards to

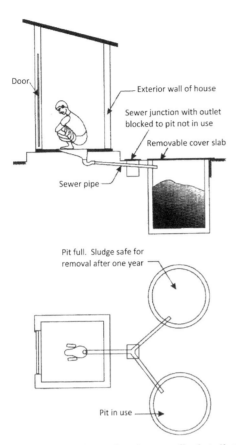

Figure 4.5 *The twin-pit pour-flush toilet*

Source: WEDC, © Rod Shaw, with adaptations by
UNICEF India

control quality, and other considerations necessary for mainstream commercial or governmental uptake, were central to their concerns.

When India's Centrally Sponsored Rural Sanitation Programme (CRSP) was launched in 1985, it adopted the twin-pit pour-flush as its regulation toilet. The pan was ceramic so as to be maximally smooth and easy to clean. And to deal with the objection that Indians would not want to use a small and claustrophobic cubicle, the 'house' was to be spacious and built with brick and mortar. These specifications made for a superior toilet – in effect, the Hyatt Regency of pour-flush facilities – but they also made the toilet extremely expensive. To be fair, to engage the Indian engineering establishment with pour-flush, on-site sanitation at all was something of a triumph, and the principle TAG focus was urban. But it was definitely a mistake to adopt this model as the standard, one-size-fits-all toilet for rural areas. The only potential household customers in the countryside were in the upper-income landowning category living in large permanent dwellings. The rural poor were excluded: even if they were motivated to want such a toilet, which they were not because no effort was made to motivate them, and even if they were given it virtually for free, as the government intended, such a structure would be out of place – even assuming a place could be found for it. This was, at best, a landowner's toilet, a chairman's toilet, a retired civil servant's toilet; not a toilet for the ordinary villager.

In the 1980s, the ideology still driving India's planning process was that the way to deal with poverty was for the government to provide free or heavily subsidized services. The idea of training up local masons to undertake sanitary construction was rejected. Instead, contractors would be hired by state and district authorities in the usual way and be given a technical blueprint to follow. There was only ever enough money in the budget to build a few Hyatt Regency toilets, so – whatever the CRSP's intention – what actually happened was that a handful of influential local figures had Grade A toilets built for them at the state's expense. Where some were constructed in the households of poorer villagers, since there was no hygiene education nor effort to sell the new behavioural idea, they were used for storage of grain or other valuables. Within a few years in any given community, little would be left to show for 'rural sanitation' except the handful of toilets in the larger houses whose owners had from the start been fully capable of paying all the costs. Some industrious and public-spirited village chairmen introduced 'village sanitation plans' and set hygienic change in motion: in India, such is the vast diversity of experience that no generalization covers every case. But overall the CRSP had a poor initial impact, and one from which it has proved difficult to recover.

The CRSP was a classic example of a supply-driven programme, in which the state adopted a technological device and set out to engineer 'sanitation for all' by constructing it millions of times over. This formula is known by irreverent Indian officials as COW – 'contractor oriented work'.[41] State budgets are used to employ contractors to build things about which the beneficiaries are not consulted, things which they may find useful or not but rarely regard as 'their own'. (Today, village and district councils, or *panchayati raj* institutions, manage local budgets, which means that they are closer to the people, but the mechanics of expenditure are similar.) No-one bothered to 'sell' rural people a new behavioural code associated with the service. Public health engineering (PHE) departments drew up plans for depositing toilets on beneficiaries and measured achievement by counting numbers built. As tends to be the case with COW, quality was often poor. Once established, however, the mould of the programme was very difficult to break: it was not in the interests of state or district officials to cut back subsidies, lay off contractors or deprive local figures of anticipated 'rewards'. Only in Midnapur, West Bengal, where the approach started from the opposite premise and was entirely based on promoting consumer demand for simpler devices made in local production centres, was the state government willing – on the basis of effective demonstration by the Ramakrishna Mission programme – to embrace a quite different approach. Here, 600,000 toilets had already been installed by 2001, when the state government put its weight into universal coverage.

Elsewhere, in much of the country, years of sanitary effort and expenditure produced little of permanence. Coverage was reported as steadily rising,[42] but without any inbuilt sustainability, it was difficult to believe the figures, which were not about sanitary behaviour or toilet use, but about numbers of constructions. In 1999, the central government relaunched the programme as the 'Total Sanitation Campaign', incorporating many lessons from Midnapur. The new goal was supposed to be to encourage toilet take-up rather than simply build facilities under COW arrangements. Local masons were to be trained, subsidies were to be reduced, cheaper versions of the pour-flush (one-pit and a non-ceramic pan) were introduced, and overall responsibility for sanitation was removed from the PHE departments and vested in local councils or *panchayats*. Rhetorically at least, the need to centre on consumer interests and popular demand was vigorously expressed. Since then, coverage has been reported in some parts of the country to have risen to dizzying heights, and some thousands of villages have recently been awarded the President's new Nirmal Gram Puraskar prize for reaching 'total sanitation'. This is undoubtedly an encouraging sign of progress, and perhaps a sanitary corner is finally being turned. However, in many parts of India there is still far to go.

A reporter from *Down to Earth*, the magazine published by the Centre for Science and Environment (CSE) in Delhi, recently went to Uttar Pradesh to see what was happening under the Total Sanitation Campaign.[43] The gap between the 70 per cent sanitation coverage claimed by district officials and the actual presence of clean and functioning facilities in the villages he visited was a chasm: only around 10 per cent of the facilities were working, and in many cases this was because the owners had spent extra money upgrading shoddy installations. When invited to offer an explanation, village landowners suggested that local authorities' costing procedures had been poor: the toilet in the approved design could not be built for the sum meant to cover it by a combination of subsidy and individual contribution – a total of 1900 rupees (US$47). Officials also trotted out the old chestnut about 'cultural resistance', but the reporter heard no-one express reluctance to use a toilet – women and the elderly expressed the reverse – as long as it was a good toilet and not constantly blocked, broken or overflowing.

Some enterprising village chairmen take matters into their own hands. Such a person is Changal Narotham Reddy, the *Sarpanch* (Chairman) of Topugunda Village in Medak District, Andhra Pradesh. (His wife became *Sarpanch* when his term was over, so his de facto occupancy of the role has been more or less continuous.) In 2003 Changal set his heart on winning the Nirmal Gram Puraskar award for 'total sanitation villages'. Not only does the *Sarpanch* of a winning village get to receive a medal and certificate from the President of India in person, and at the Presidential Residence in Delhi, but there is also a grant of 200,000 rupees (around US$5,000, a substantial sum in Indian village terms) for village development. By a process of leadership and persuasion, backed up with state and central subsidies (2100 rupees, or US$52, for a toilet, extra for a bathroom), Changal Reddy succeeded over 18 months in having every household in his village install a pour-flush toilet. The poorest families were given extra help by the village council. Lanes and drains were cleaned, and a special small child's toilet installed at the day-care centre (*anganwadi*).

In 2006, the village was duly adjudicated 'TS' (total sanitation), and on 23 March 2006 Changal Reddy collected the award in Delhi at the *Rashtrapathi Bhavan*: to have attended the President at his home was one of the most important things to have happened in his life. The village has since received visitors from far and wide, including from the states of Maharashtra and Karnataka and even from South Africa, asking Changal for his toiletization and 'clean village' recipe. To his pride, the village has become a tourist attraction. The approving state government also chipped in with extra funds, with which he has paved all the village roads.

Topugunda was one of ten villages in Andhra Pradesh to achieve TS status in 2006. That year, there were 775 TS awards nationwide, up from 35 the previous year, and in 2007 there was a further great leap forward – to 4959. However, with around 229,000 villages in the country, significant challenges remain, especially among poorer villages, villages in tribal areas, and villages with large numbers of landless or *dalit* (scheduled caste) inhabitants. Changal Reddy is convinced that the slowness with which communities that do have the resources take up sanitation is due to the political and official silence on the topic:

> In all the major forums, in the State Assembly and in the Parliament, they never mention this topic. They don't even publicize the Nirmal Gram Puraskar award. They talk about roads, electricity, waterworks, education and jobs, but they never discuss sanitation. Only one member of the State Assembly has showed any real interest. But sanitation is just as important as the other subjects. It should be a regular item on the agendas of decision-making bodies.[44]

For that to happen, the public distaste which continues to obscure the subject needs to be actively challenged.

Although the presidential prize has generated much more momentum around sanitation than there used to be, in too many parts of the country the

TS Campaign risks performing like many other rural development initiatives in India: less an opportunity for empowering local people than a source of income for contractors.[45] To move from supply- to demand-driven service delivery approaches in settings such as these requires a radical upheaval in mindsets and power relations. Somehow 'good governance' – the parlance used to describe such transformations in international donor circles – doesn't quite capture what is needed. How do you bring about a multiplication of public-spirited local politicians prepared to take village sanitation and long-term behavioural change meaningfully by the horns – a challenge not only in India but virtually everywhere in the developing world? A national prize incentive is one way of 'spreading the word' and attracting the attention of local councillors and administrators, and this example is one that could be emulated elsewhere. But other taboo-breaking initiatives are also required to make a real difference.

Some observers put their trust in generating more demand – and in the market lanes of old Indian cities, there are today more small shops advertising 'hindware' or toilet pans than in the past. But it will take time for consumer power to effect a sanitary revolution in countries with huge populations below the poverty line. It will probably come via accelerating desire for toilets alongside other modern gadgets such as televisions and mobile phones. But among the less educated, it needs a hefty push. Until officials and politicians lose their reluctance to talk openly about lavatorial matters, out in the small towns and villages of Asia, genuine progress towards total sanitation will still only crawl forward, more in step with the bullock cart than with the railway express.

The article in *Down to Earth* lamenting the narrow range of technological choices available under the TS Campaign in villages in Uttar Pradesh quoted an interview with one of the world's leading sanitary activists, Uno Winblad of Sweden.[46] Winblad had expressed regret that, during the 1980s, the entire preoccupation with low-cost sanitation in Asia had settled on the pour-flush, courtesy of forceful representations from the World Bank's Technical Advisory Group (TAG): 'If someone tried to come up with alternatives, for example, dry, non-polluting systems entirely above ground, this was treated as heresy.' This interview was undertaken shortly before the first International Conference on Ecological Sanitation, held in Nanning, China, in 2001.

For many years Winblad had been expounding the virtues of ecological sanitation (ecosan) – essentially the same approach and on similar grounds as that advocated by the 'dry conservancy' enthusiasts of Victorian times – almost as a lone voice in the wilderness, certainly an eccentric one as far as the TAG was concerned. Dry toilets, such as Vietnam's 'double vault composting', and less well-known examples such as the long-drop systems used in urban Central Asia, had important virtues which Winblad believed were being ignored. He was frustrated with the worldwide fixation on water-flushing for sanitary progress, especially in places that were water-short, or where hydrogeology or settlement patterns made the pour-flush unsuitable. Over time, Winblad's doggedness and missionary zeal won him a following among other Swedish and Norwegian enthusiasts, as well as from Canadian, German, Chinese and Japanese support-ers, some NGOs, and a few influential voices in international public health, notably Steven Esrey of UNICEF and Ingvar Andersson of UNDP. The inter-national conference at Nanning, which Winblad brought to fruition almost single-handedly, was the moment at which the international seal of approval was attached to 'ecological sanitation', and even sceptical public health officials were pushed into giving 'ecosan' a hearing.

In the period since sewers became established as *the* way to deal with excreta removal from households and neighbourhoods, more had been learned about the science of composting and how to make dry conservancy work. The Chinese, motivated by agricultural application of human faeces, were the pioneers, but their methods were only brought to the West in the early 20th century – too late to be of use in the first 19th century public contest between 'wet' and 'dry' disposal systems.[47] Experiments with faecal composting – with variables such as temperature, alkalinity, radiation, deprivation of oxygen – continued down the years in various scientific laboratories in different parts of the world. When Winblad, an architect and urban planner, was posted to Ethiopia in the 1960s, he faced problems of sanitation in Addis Ababa which defied all available conventional solutions. The alternatives he sought needed to be unaffected by rainwater flooding and overflowing pits, and not to require underground networks of pipes. His subsequent support for 'sanitation without water' reverted to the idea of composting, which had been effectively eclipsed from low-cost sanitation menus in the Water and Sanitation Decade drive.

As it evolved from this point onwards, the 'ecological sanitation' thesis was roughly as follows. The sanitary crisis and high rates of diarrhoeal disease in many urban (and rural) areas of developing countries were the product of faecal contact caused by lack of effective excreta disposal. Flushing systems were unaffordable for most poorer areas, including urban slums, and ended up contributing to environmental pollution. In many countries, agriculture was still an important means of livelihood and people were often undernourished. The

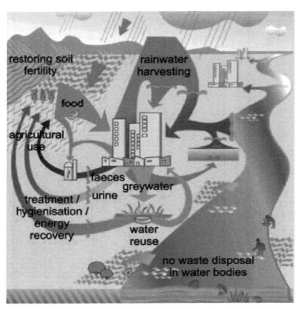

Figure 4.6 *The advantages of ecological sanitation*

The diagram shows that ecological sanitation is an approach, not a particular technology. At every stage in its use, water falling as rain or extracted from streams can be reused or recycled so as to derive the maximum benefit from the water itself and from its use in sanitation and waste transport, and to reduce to a minimum the discharge of polluting material and the presence of pathogens in the environment.

Source: Christine Werner, GTZ, 2007

loss of a free supply of nutrients for growing food and other crops, and its replacement by an unaffordable manufactured alternative in the shape of artificial fertilizer, was inappropriate and wasteful. On top of the nutritional loss, it was perverse to squander water that was scarce, precious and distant from the home on flushing, and even more so on transporting excreta and dumping it in waterways. Still further objections applied in terrain where toilet pits quickly became waterlogged and pollution spread into wells or streams.

The 'ecological' solution was to contain what Winblad refused to call 'wastes', and instead called 'resources', until the faecal content was sanitized and safe, thus averting the risk of spreading contaminated matter in surrounding water and soil, and to reuse the composted residues as dry manure. Thus the loop between what humans take in and what they put out would be hygienically and scientifically 'closed' (Figure 4.6).[48] This ecologically tidy system was promoted by ecosan warriors as an alternative both to standard and improved pits ('drop and store') and to the water-wasting process of 'flush and discharge'.[49]

Following the first Earth Summit in 1992, growing worldwide recognition that water was a valuable resource under increasing pressure, and of the mathematics of water stress and water scarcity in many of the developing world's cities, gave a major fillip to the ecosan cause. Environmental issues were moving smartly up the international agenda, and concern about water was helping to drag up even that least mentioned of topics, sanitation, given the increasing threats of pollution and epidemic disease posed by the randomly distributed faecal output of more than 6 billion people. The wasteful economics of sewerage were becoming more conspicuous. To utilize 15,000 litres of costly fresh water to flush away 35 kilos of shit and 500 litres of urine – the average generated per person per year – could not possibly make sense, especially when cost–benefit and water-use efficiency were now being given their due.[50] In many towns and cities of the developing world – there were already 600 in China alone – water supplies were under such stress that there was flow in household pipes only for a few hours every day. In poor and water-short cities in Central Asia, such as Sana'a, Yemen, and Kabul, where traditional 'dry' vault systems have been used for centuries, what could possibly be gained by attempting a replacement with water-borne arrangements?

Examined without reference to aesthetic considerations, conventional water-borne sewerage as set up and managed in the contemporary world appeared an extravagant folly – at least, for large parts of it. Via WCs, tiny amounts of extremely pathogenic material are added to large quantities of drinkable water, which is then flushed away to be thoroughly cleaned to take the pathogens out again. The environment the pathogens have polluted, courtesy of their water transport, may also have to be cleaned at very considerable expense. When all of this has been done, the water is then sent back down the pipes to start the profligate cycle over again. No-one would propose this as *the* universal solution for excreta disposal if we were starting again from scratch today. By comparison, the composting toilet, requiring no water and producing fertilizer and, potentially, biogas for cooking and lighting, appears a miracle of sanity. Its advocates regard it not only as the answer to sanitation in low-income areas of the developing world, but as a better toilet technology for virtually everywhere. However, this is hotly contested by ecosan critics, who protest that initial investments tend to be high and that cultural – and consumer – preferences are not in ecosan's favour. Ecosan supporters insist that the technology is no more intrinsically expensive than alternatives. In theory, the product has agricultural value and the system reduces treatment costs to zero; but so far, such economies have yet to be realized on any scale. In 19th-century Europe, 'dry conservancy' on a mass basis never proved economically viable, as noted in Chapter 1. Today's advocates reply that the politics and economics of sanitation systems, as of transport and energy in a pressurized world, are rapidly changing in favour of nutrient recycling and more sustainable solutions.

Expense of construction of the ecological toilet and considerations of space are important practical considerations. But leaving those to one side, what of the ecosan toilet itself? Like the dunny or outhouse of early 20th-century folk literature, the earth closet was ridiculed; its modern improved incarnation, the

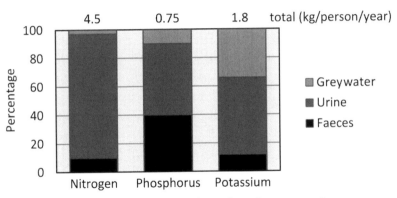

Figure 4.7 *Nutrients from human waste*

Source: R. Otterpohl (2001) 'Black, brown, yellow, grey: The new colours of sanitation', *Water*, vol 21, pp37–41; cited in Duncan Mara et al (2007) 'Selection of sustainable sanitation arrangements', *Water Policy*, vol 9, IWA Publishing, pp305–318

VIP, has not had the smoothest of rides either. Can any 'dry' toilet occupy a position of equivalent status and desirability to that of any 'wet' toilet, if consumer demand is put at the centre of the frame? And the ecosan toilet – at least as far as ecosan purists are concerned – has an extra dimension redolent of toilet arrangements of civilizations gone by, requiring considerable personal restraint. This is 'urine diversion' (UD). In the UD toilet, there are two compartments, and the user has to position himself or – more problematically – herself to commit the different excreta to the separate compartments in the pan. No liquid should be allowed to descend into the shit bit – although for cleaning or by mistake, some spillage is inevitable. To replace the odour removal role of a water-seal, a handful of ash, lime, sawdust or other suitable dry material (often soil) is deposited on top of the pile.

Ecological sanitarians argue about whether urine diversion, given human undependability, is intrinsic to their cause, but its recycling and reuse rationale is sound. The solid matter on its own is small in volume, and contains virtually all of the pathogens in excreta. Urine is nearly sterile and contains most of the nutrients needed for agriculture – over 80 per cent of the nitrogen, and half the potassium and phosphates (Figure 4.7).[51] Thus storage problems are reduced and fertilizing benefits increased by separate collection. For users, an important aspect is that the separation, even without a handful of lime or ash to flush, reduces odour. The combination of the two is what really makes a toilet-house stink in an overpowering manner.

There are problems associated with selling existing toilet users pedestals or squat plates with urine diversion which, when the imagination is applied, need little exposition – though one Swedish commentator proposes the use of behavioural psychologists to examine the question.[52] Apart from the question of user accommodation, there are issues to do with emptying the vaults and bottles, and use (or disposal) of the end products. Here is a context where 'cultural resistance' does indeed seem to play the part it is too often unfairly ascribed in relation to toilets generally. The word 'faecophobic' is used to describe societies unwilling to touch faeces wet or dry, and it will indeed take some persuading to get self-respecting Madagascans, Indians, Ugandans, Mexicans or many other peoples to handle excreta in no matter how inoffensive a form. However, in China and some other parts of East Asia, this presents no problem. As a result, ecosan programmes were first promoted in this region, with the Vietnamese DVC and certain Japanese appliances in the middle of the frame.

China is known for its faecophile attitudes and a lack of taboos concerning defecation. In the mid-1990s, when Uno Winblad first began to work in China, raw excreta still lay about in the countryside virtually everywhere, public lavatories were a fright, and excreta disposal, whether in rural or urban areas, was often disgusting. Nearly half the children still suffered from ascaris (roundworm) infestation, and other menacing excreta-domiciled parasites and pathogens played havoc with millions of people's health on a daily basis. Most households had a toilet of sorts, but by no stretch of the imagination could more than a tiny minority be described as 'improved'. As Winblad described, and as happens in many semi-urban areas in the developing world, 'Many villagers are building new modern houses, but sanitation is being left behind. In China, a household can spend money on a luxury house, with mirrors in the ceiling and marble on the floor, but the toilet is still an open stinking pit in the backyard.' The task of upgrading sanitation in poor rural areas of China is truly Herculean.

With assistance from Sida (Swedish International Development Cooperation Agency) and UNICEF, an 'ecological sanitation' trial was started in Yongning County in southern Guanxi Province in 1998. The idea was to create 'eco-villages' where renovation of sanitary facilities was combined with other improvements, including the installation of reliable electric and phone systems. Among other activities – road paving, tree-planting, corralling of livestock, installation of biogas plants – every household in every eco-village had to install a UD toilet. This meant building a tiled bathroom compartment inside the house, next to an exterior wall. This compartment included a raised double composting chamber with an outlet in the wall, on whose tiled top was fixed a plastic-moulded, two-ended squat plate with a lid. A bucket of ash or lime was kept beside it for the flush. When one chamber was full, it was closed for a year. Once the pathogens were truly dead and the content turned into an entirely unobjectionable friable substance, it could be extracted via a hatch in the exterior wall and used as a soil-filler.

Clean, compact and cheap (around US$35), the UD toilet was regarded by Yongning villagers as a vast improvement on their previous arrangements, which no-one would have dreamed of situating within the house. One model could even be installed upstairs, a moveable pipe leading down the exterior wall into one of the composting chambers. The toilet provided, in addition to the joys of a proper bathroom, an economic benefit. Outside, a plastic pipe led fresh urine from the front end of the toilet into a large blue bottle, ready to be diluted and used to fertilize the vegetable garden. By 2003, when the project ended, several hundred thousand UD toilets had been installed in Yongning County, and the approach had already spread to 17 provinces and 685,000 households altogether.[53] The success of this programme attracted many converts to ecosan and helped bring on board resources and the kind of authoritative backing that has enabled the concept of nutrient recycling to enter the mainstream sanitation portfolio.

Across the world in Central America, ecological sanitation has a somewhat longer history. Inspired by Ivan Illich, the celebrated advocate of alternative thinking who lived in Mexico for many years, architect César Añorve decided in the early 1980s to dedicate himself to *alternativos* in the field of sanitation. Water conservation and recycling were his principal interests and he set up an NGO, the Center for Innovation in Alternative Technologies (CITA), in Cuernavaca. Añorve imported the Vietnamese double-vault composting toilet and adapted it to the Mexican context by designing a UD pedestal bowl.[54] Made of moulded fibreglass, the toilets were fabricated at his workshop; they used ash or lime for the flush, and were described as 'DVCs with alkaline desiccation'. Añorve went on to train others in their manufacture and to sell and export his ecosan-ware around the region – and further afield, to the NGO Mvula Trust in South Africa, for example.[55] Although there was potential for recycling in agriculture, the toilet's main selling-points in the Mexican context were water shortage, the dreadful state of existing facilities and demand for a toilet which was pleasant enough to install in the house. By definition, however, this only applied to people of a certain income and housing level. Although Añorve's work was important in a pioneering sense, and became the basis for the spread of ecosan ideas in the region, the scale of his operation was very small, and the costs of his toilet placed it beyond the reach of poorer customers.

Today, in neighbouring countries, the principles of ecosan are widely known. In El Salvador, for example, the Ministry of Public Health and Social Assistance

(MSPAS) has recently been conducting a research project on different kinds of eco-toilets. In the past, there were some efforts to promote improved pits to the 80 per cent of rural inhabitants for whom flushing toilets are impracticable, but because there was no accompanying health education, the vast majority were never used. In 1996, the MSPAS brought in an expert in participatory learning from Mexico and trained up a cadre of hygiene educators. They have also borrowed from educational materials developed in Honduras by UNICEF and from experiments in the region with ecosan toilets. In the ten years since hygiene education has been underway, Rigoberta Rivera, a senior officer in MSPAS, has noticed behaviour beginning to change: 'When a toilet is damaged, people now go and seek assistance to repair it. They have begun to appreciate the value of *letrinas*. But unless a programme offering them is available, demand does not get expressed. So we still have much to do.'

At the edge of the Pacific Ocean, perched on a hillside with spectacular views over a coast-fringed luxury estate, nestles a new and impoverished community called Los Angeles. The local municipality confiscated this land from its owner in lieu of taxes some years back. A few years ago, 55 impoverished families who lost their homes in a 2001 earthquake were given permission by the mayor's office to settle here, against the protests of villa-owners in the plain below. The settlers had to struggle hard to make a new life. The road to their eyrie is a track which has to be painstakingly descended and reclimbed in order to go anywhere at all – to shop, to work or to do the laundry at the river. There is no employment here, no water supply nor school, but for the inhabitants of Los Angeles this is a wonderful improvement over nothing. The local Rotary helped with the housing plan and materials, and gradually the community is taking shape.

Los Angeles has been chosen by the Ministry of Public Health to try out the *'letrina solar'*, one of three models of ecosan toilets they are field-testing around the country.[56] The toilet has some aesthetic and structural variations (the most solid version is earthquake-proof, for example), but the basic appearance is the same. Brick cabins familiar to toilet tourists in the developing world are dotted about the Los Angeles plots, looking well tended or sad in the same way that any housing does in a community where some inhabitants are doing well and others have given up. Some Los Angeles matrons are energetically house-proud, and happy to discuss the performance of their *letrinas*.

What is striking about these toilets, as with many on-site facilities, is that they require work and pose problems that those with standard flush toilets in industrialized environments cannot begin to appreciate. The solar toilet used in El Salvador is UD, meaning that the composting chamber can be kept relatively

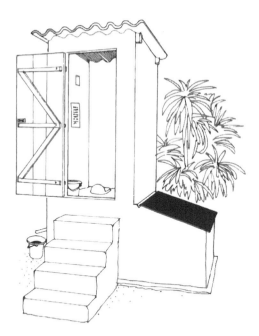

Figure 4.8 *A solar ecosan toilet*

This eco-toilet with solar-heated processing chamber was designed by Uno Winblad and Duong Trong Phi of the Nha Trang Pasteur Institute project in Duc province, Vietnam, in 1996. Similar models are in use in many locations, including El Salvador (see text).

Source: Mayling Simpson-Hébert and Uno Winblad (eds) (2004) *Ecological Sanitation*, revised edition, Stockholm Environmental Institute, p66

compact. The shallow pit is situated so that only a part of it is under the cabin; the rest is off-set and protrudes outside (Figure 4.8). Its top, slanting from the cabin wall, is covered by a black hatch positioned to consume heat from the sun. Once a week the hatch has to be opened and the content covered in an extra layer of ash or lime and shovelled into one of the two compartments. One compartment is expected to fill in around 45 days, but will do so faster when the family is large. Once a further 45 days have passed, the first chamber has to be emptied and the contents buried before it is again ready for use. So management of faecal matter with a *letrina solar* – even where there is no intended use of the product as an agricultural manure – is relatively labour-intensive. Needless to say, the weekly maintenance work is all done by women. In little Los Angeles, where the sun is obscured for part of the day by trees and hill-tops, research has found that it cannot be relied on to cook parasites to death. So despite the women's efforts to maintain their toilets in the best condition, the health safety of the system is still not guaranteed.

People in Los Angeles are pleased to have a toilet, especially one which does not smell: 'When someone comes and asks if they can use the bathroom we are in a position to offer something clean,' says Juana Antonia Alvarado, one of the local mothers. But she does have problems. One is the cost of lime: she needs to put half a bag on the ordure every week, which at US$0.40 a time she cannot afford. Also, it pours with rain frequently in El Salvador, and in the 18 months since the toilet was built, the black hatch covering her pit has deteriorated. When it rains, water leaks in and creates condensation, allowing bacteria, worms and parasites to flourish. There is no service for Juana Antonia to call upon to mend her toilet, and problems in other areas of her life are pressing enough to keep her fully occupied. Her husband has recently lost his job and she has to support the family by occasional work as a bartender. Her children do not go to school because she cannot afford the uniforms. Such a family has difficulty finding enough money for food, let alone to upgrade their housing. Without the support of the local health programme, and without subsidies, very few residents of Los Angeles would enjoy the use of a toilet.

The other forms of ecosan toilet explored for use in El Salvador are the modified pit, which is much like a VIP with a vent-pipe and a UD system, suitable for low water-table areas, and the dry composting family double-vault suitable for high water-table areas or places so rocky it is hard to dig, with two raised compartments placed on top of the ground like the Vietnamese DVC. In all these systems, as in Central America generally, ecosan is mostly taken up because of its freedom from smell, the lack of water availability for flushing, and the smaller amount of space needed for faeces containment only. Agricultural use is so far negligible, so the much-vaunted loop from human output to human input has not been closed.

Back in Africa, a number of ecosan toilets have established themselves in the armoury of low-cost sanitation in recent years, mostly but not exclusively in rural areas. Many were originally proposed by Peter Morgan, inventor of the Blair, to whom water scarcity, environmental considerations and the attraction of closed loop thinking were inspirations. One such commodity is known as a 'skyloo' because it is raised up on the ground over its own chamber, instead of a pit with a hole at ground level.[57] This toilet, also a solar and DVC, is useful where the soil is rocky and hard to dig. Some types of skyloo use removable bucket containers whose contents are put into a secondary composter and retained for 12 months (Figure 4.9).[58] Morgan's more basic approach is to dig a relatively shallow pit (one metre deep) to use for a year, then shift the above-ground contraption and place it on top of an identical pit. The only necessary items are a ring plate and squat plate, both cast in concrete to place over the hole; and a cabin made of mud, wattle and thatch. The original pit can then be covered over and its contents left to compost for a year. This has been designated the 'fossa alterna'. Even simpler is the 'arborloo': when one pit is full, it is closed over with earth and a tree planted in it.[59] The superstructure simply travels from pit to new pit whenever necessary.

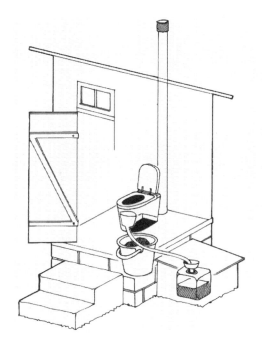

Figure 4.9 *The skyloo in Zimbabwe*

This version of the skyloo has a urine-capturing device in the pedestal which diverts it through a pipe into a container. Faeces drop directly into the bucket in the vault, and users add dry soil and wood ash after every visit.

Source: Mayling Simpson-Hébert and Uno Winblad (eds) (2004) *Ecological Sanitation,* revised edition, Stockholm Environmental Institute, p43

When the manuals for building and managing these pit toilets are examined, the impressive feature is the huge number of variations: depth of pit, positioning of vent or pit-covers, pedestal or plate, *pukka* or *katcha* (Indian slang for solid brick or natural mud and wattle construction), dependency on ash, lime or water for smell-reduction, dependency on temperature, alkalinity, UV radiation or desiccation for rendering shit into compost, and UD or non-UD. Like any new idea that enters the development lexicon, ecological sanitation has been conceptually and practically subdivided into many different incarnations since the day when the pioneers first set out the virtues of closed loop thinking. This is a sign that ecosan has been removed from its backwater and landed squarely in the mainstream. In some water-short corners of the world, the ecological approach is being 'rediscovered': in Kabul, for example, 48,300 traditional double-vault toilets were renovated and improved in the late 1990s, with larger chambers, urine separation, vent-pipes and sealable doors on their emptying holes to prevent raw faeces leaking into the street.[60] In some municipalities in South Africa and Sweden, ecological sanitation has begun to be incorporated into housing estates and apartment blocks, showing that this is not just a sanitary technology for the poor and difficult-to-sewer, but can be a respectable technology for certain industrialized settings.

There was a moment, especially among its Scandinavian, Dutch and German supporters, when ecosan became *the* approach for the early 21st-century sanitary revolution – the new holy grail of sanitation. Water-borne sanitation was rightly excoriated for its squandering of valuable water supplies, the environmental pollution it causes and the loss to agriculture of recyclable nutrients. Progressive depletion of the world's scanty phosphorus deposits means that this readily available alternative source is daily gaining in economic significance. Important principles which had disappeared since 'wet' beat out 'dry' in 19th-century Europe have been reintroduced into the science of excreta management, and are undoubtedly of value. But in all branches of social improvement, there is no 'one-size-fits-all' solution, and this applies as much to ecological sanitation as to anything else. Most people used to a water-seal porcelain toilet which accepts both forms of waste plus cleansing materials and disposes of them with infinite ease will be difficult to persuade that UD and dry systems are superior, whatever their ecological merits. Aesthetics, convenience and pleasantness are unchallengeable winners in environments economically able to uphold the social and consumer status of the in-house bathroom and WC. The water-flushed toilet, whether with a handle and tank, or with a bucket at the side, is not losing serious ground to dry conservancy or UD, nor is it likely to. Even where water is not used for flushing, some low-income communities where sanitation is being introduced prefer a VIP, or an improved pit toilet with a sanplat, to a composting UD – if given the choice.[61] Like all of its predecessors, ecological sanitation is in no sense an exclusive solution.

Dry and composting toilets ought to be advocated wherever they are useful and affordable. This will principally mean rural areas where there is sufficient space for them, although there are also some towns and cities – including Sana'a and Kabul in Central Asia, some South African municipalities, urban Mexico, El Salvador and Guatemala, and China – where they are being successfully used, and there are no doubt others where housing is spacious enough and attitudes open enough for them to be considered. Even in faecophobic India, there are now manufacturers developing UD squat-toilet plates, and the market will undoubtedly develop here and elsewhere as time goes on. But most of the evidence is that, given a choice between 'wet' and 'dry', people new to sanitary ware, especially in faecophobic societies, tend to prefer 'wet'. If consumer demand is to be the driver for sanitary take-up in untoileted areas of the world today, water is not going to be banned from the pan or U-bend any time soon.

The great extension of the on-site menu for toilets has been an important breakthrough for sanitary advance. But without hygiene education and demand cultivation, no toilet device, wet, dry or any combination thereof, will enjoy rapid take-up. The role of persuasion in sanitary change is the next subject for consideration.

5

Selling Sanitation to New Users

Previous page: Girls' toilet block in a village school in Casamance, Senegal. The blocks are designed so that girls enter and then turn in to the toilet – more dignified than opening the door to the world. There are no roofs: during seasonal torrential rains when school is closed, the blocks receive a thorough drenching.

Source: UNICEF Senegal/Idée Casamance

If the Water Decade proved one thing, it was that, to bring about the sanitary revolution, the engineers would have to concede pride of place to the agents of behaviour change. The technicians had led the way in identifying suitable technologies for non-sewerable parts of the world, compiled manuals for their construction and worked with technical institutions around the world to secure their professional recognition within the civil engineering establishment. But hardware was only one part of the equation. As one commentator put it, even the best manuals cannot teach an engineer to be sensitive to the needs of an impoverished community.[1] Just like the proverbial horse that is taken to water but cannot be made to drink, people in the urban and rural developing world could be presented with immaculately engineered toilet facilities, but not be made to use them. Unless they lived in circumstances which predisposed them towards toilets, why would they? To be fair, engineers are not expected to be good at behavioural change: what they are trained to do is 'fix'. More sensitivity to lavatorial demand (or its creation) and to behavioural change was the new leitmotif. 'Software' and motivation – how to sell the toilet habit convincingly and durably to new users – became the principal preoccupation of toilet missionaries.

When the sanitarians first tuned in to 'software', the emphasis was on the incorporation of health and hygiene education into whatever else they were doing. 'Faecal perils' (a literal translation from the French term) were not sufficiently understood among the unschooled. 'Why are the flies sinners in our district?' asked a villager in a programme in Imo State, eastern Nigeria. Until people appreciated the disease-laden properties of the tiniest morsel of shit, even when invisible to the naked eye, they could not be expected to embrace toilets, wash their hands and do the necessary to break the faecal–oral pathogen route. So this Nigerian programme, launched in 1981, allocated large amounts of management effort, personnel and time to health education.[2] The linchpin was the 'village-based worker' (VBW), a community volunteer trained to inform, influence and provide help to neighbours, very much in line with the 'strategy for primary health care' recently adopted by UNICEF and WHO. These VBWs, young men and women nominated for training by local chiefs and councils, undertook a course lasting several weeks. They learned the elements of maternal and infant care, safe excreta disposal, personal and domestic cleanliness, diarrhoeal disease prevention and treatment, breastfeeding and nutrition, maintenance of handpumps, and production of VIP toilet components – the whole gamut of public health in its broadest interpretation.

A study conducted by the London School of Hygiene and Tropical Medicine (LSHTM) was carried out in Imo State as an integral part of the programme:

this was the period in which Richard Feachem and his colleagues were trying to identify the precise impacts of providing access to clean water and sanitation, and a health-monitoring component designed by the LSHTM was included from the outset. The main health benefit achieved in Imo turned out to be a drop in malnutrition in the under-threes.[3] People were also taking more care to keep household drinking water clean; but they had learned this from the project personnel, not from the VBWs. The latter's performance in altering local behaviours – particularly with regard to installing and using VIP toilets – did not seem to justify the considerable investment in their training. Thus from 1986, the programme began to rely less on VBWs and look for other ways to motivate people to adopt sanitation. Meanwhile, versions of the programme were exported to several other states in Nigeria and its design continued to evolve. One of the lessons demonstrated by the pilot in Imo State was that education was not the quintessential missing link between water, sanitation and better health – not, at least, as designed and delivered in this location. Having recognized that offering appropriate 'hardware' was not the beginning and end of the story, experts now discovered that identifying the right 'software' to 'sell' people sanitary change was not easy either.

Although the operational twinning of sanitation with water supplies has often been unhelpful to sanitation, especially in contexts where water is not implicit in any part of the excretory performance from cleansing to flushing to waste transport, one innovation of the 1990s was to try and use the linkage with water as a lever. There might be no demand for sanitation in settings where fresh air was abundant and the flies' sins were not excessive, but that did not affect demand for new water systems. Water might be distant or in short supply, especially in the dry season, and people might be desperate for something closer and more reliable than a seasonal spring or stream. Gauging demand was a new science, and assumptions were often too freely made about the strength of local desire for community water installations – notions exploded by the frequent abandonment of pumps or taps needing only minor repairs. But where there was real demand for water, this could be used as a negotiating chip. As in Imo, new-style 'integrated' programmes started with training for hygiene and health promoters, progressed to toilet persuasion, and only when a certain number of VIPs or pour-flushes had been installed would the drilling rig arrive or the standpipes be connected. By this stage, a local 'watsan' committee would have been set up, levies for connections and maintenance agreed, and theoretically the community would be on its way to a sustainably clean and safely watered future.

Such programmes, with their emphasis on community involvement and expression of demand, were a radical improvement on their predecessors. Courtesy of 'integration', the hygiene promoters could propose the use of the new water supply for new hygiene behaviour – especially washing hands to remove itinerant faecal particles, keeping drinking water in a separate container and keeping the entire domestic environment clean. There was also the incentive of better personal grooming. When asked what he thought was the most important benefit of the Imo State programme, a local schoolteacher thought for a moment, and then said that his wife was now always cleanly and freshly dressed.[4] As emphasized earlier, few people anywhere rate the health virtues of toilets as reasons for their installation (Table 5.1). So hygiene education might lead to certain kinds of behavioural change, such as washing hands or using soap, improved appearance and better dress, but not necessarily to the construction or regular use of a toilet – an altogether more expensive and dramatic lifestyle change.

Even when people fully understood that flies were invariably sinners and often spread disease, most of the builders of toilets in Imo State built them because they were told to do so in order to obtain other benefits. With the exception of the local *Eze* or king, whose palace compound had been suitably enlarged, once the programme teams moved on, people fell back to their old habits. The VBWs, mostly young men and women without status or influence, stopped visiting compounds to recite the benefits of sanitary behaviour. The large solid cabins dotted about the landscape mostly became monuments to the zeal of those insisting on their construction, and conveniences for visiting dignitaries and project personnel.

In more densely settled communities, people might genuinely regard toilets they had been obligated to build positively and use them to begin with. But if toilet pits later became full and could not be emptied, or got broken and could not be repaired, then what to do? Somehow the software for that was neglected. The programme energy that went into software for *water supply* programme components – setting up management committees for water-points, making sure women participated in site choice, training handpump caretakers, ensuring that there were supply chains for handpump washers and other spare parts – rarely went into software for sanitation. The local watsan committees might be expected to promote sanitation as well, but their members were not usually given the means or training to do so. There were no pit-toilet emptying services equivalent to handpump repair systems, no toilet shops or supply chains for replacement parts or plats. Why not? Because sanitation was almost invariably

Table 5.1 *Stated benefits of improved sanitation from the household and public perspectives*

Household Perspective	Society/Public Perspective
Increased comfort	Reduced excreta-related disease burden
Increased privacy	(morbidity and mortality), leading to reduced
Increased convenience	public healthcare costs and increased economic
Increased safety for women and children	productivity
Personal dignity and social status	Increased attendance by girls at school, leading
Being modern or more urbanized	to broad development gains associated with
Cleanliness	female education
Lack of smell and flies	Reduced contamination of groundwater and
Less embarrassment with visitors	surface water sources
Reduced illness and accidents	Reduced environmental damage to ecosystems
Reduced conflict with neighbours	Increased safety of agricultural and food products
Good health in a broad cultural sense, often	leading to more export
linked to avoidance of disgusting matter,	Nutrient recovery; reduced waste generation and
especially faeces	disposal costs (for ecological sanitation)
Increased property value and rental income	Cleaner neighbourhoods
Eased restricted mobility due to illness and	Less smell and flies in public places
old age	More tourism
Manure for crops and reduced fertilizer costs	National or community pride
(ecological sanitation)	

Source: Marion W. Jenkins and Steven Sugden (2006) *Rethinking Sanitation: Lessons and Innovation for Sustainability and Success in the New Millennium*, UN Human Development Report, Occasional Paper 2006/27, New York, p3

water supply's poor relation, an incidental component of programmes delivering water supply systems instead of the other way round. The failure to make sanitation the cutting edge and the context for the whole WASH (water, sanitation, hygiene) package had a lot to do with the Great Distaste still keeping the subject under wraps.

What was needed was a complete shift in perception. The word 'sanitation' was too often narrowly perceived as toilets and nothing else, whereas in fact a sanitary environment is a clean environment. And once people began to reduce the emphasis on 'toilet' and start emphasizing 'clean', much else began to shift. Virtually everybody – not just the wife of the Nigerian schoolteacher – wants to be clean. Use of a toilet to avoid dirtying the environment needed to become an integral part of the innate desire for personal and environmental cleanliness and wellbeing.

As usual, it needed people in leadership positions to run with this idea. Such a person was the local *Bupati* or district head of West Lombok, Indonesia, who became known as 'the latrine *Bupati*'. In 1993, after a successful immunization campaign did little to reduce the lamentable rate of infant mortality in his district, the *Bupati* became convinced that the problem was largely attributable to the unsanitary environment. Toilet coverage in West Lombok was the country's lowest at 8 per cent. At a public meeting, he challenged the assembled representatives to construct 20,000 family latrines. His campaign was backed by UNICEF, which offered a subsidy of US$12 per toilet to help get things moving.[5] Water-well installation was also part of the package – a useful bait and an essential aid to cleanliness. The *Bupati*'s stroke of genius was to invoke religion and bring the local imams onside – as had been done for immunization. Thus was launched the 'Clean Friday' movement, whereby every week at Friday prayers, the people of West Lombok listened to injunctions to convert to sanitary behaviour, and went home to put them into practice. The local chapter of the Indonesian women's movement, the PKK, set up production centres for pour-flush polished bowls and concrete ring pit-linings like those in West Bengal, organized the training of local youth in these new employment opportunities, and drew up lists of candidates for assistance with installation.

In spite of the fact that previous efforts to introduce sanitation into the island of Lombok had been very discouraging, not only were 20,000 toilets built within months, but by the end of 1994, toilet coverage in West Lombok district was almost universal. This success was credited to the pressure exerted by women on their husbands, and on the emphasis on hygiene as part of religious duty. The 'latrine *Bupati*' made sure via the local imams that people without pit toilets would not be allowed to marry, nor to travel to Mecca for the Haj. So celebrated was the *Bupati*'s achievement that he was summoned by President Suharto to explain his ideas. Clean Friday was then launched nationally. Islamic leaders throughout the country were asked to associate the day of prayer with activities to promote healthy and hygienic living. This encompassed not only the use of sanitary toilets, but hand-washing, keeping drinking water free of germs and proper garbage disposal. To facilitate 'clean living', water supplies and toilets were to be installed not only in houses but in all public places such as schools and places of worship.

Although the Clean Friday movement made some progress nationally in the following years, no other district managed to reach the heights of toilet coverage and health monitoring of West Lombok: a forceful promoter with political push and solid religious endorsement makes a huge difference. In West Lombok, meanwhile, noticeable declines in respiratory infections, skin problems and diarrhoeal diseases were recorded.[6]

Clean Friday was also targeted at men. In the new generation of water and sanitation programmes, the heightened emphasis on health and hygiene education was very much associated with the recognition of women's roles in the provision of water and use of sanitation. This was another outcome of the Water

Decade, in which activists on behalf of women's watery concerns had been very effective. But sometimes, the new focus on women meant that men's household and community roles were neglected. Yes, women were responsible for everything domestic, including water collection, management of children's excretory behaviour and disposal of their faeces, laundry, washing up, food preparation, and hygiene in the home. So putting a case to them based on child wellbeing and diarrhoeal disease reduction might make a strong impression. Where women were in positions of responsibility as a group – as with the PKK in Indonesia – and were charged with spreading sanitation, progress was better guaranteed. But in many traditional societies women's ideas and opinions on matters beyond the purely domestic do not carry clout: they have little or no say in decisions about the allocation of household resources and expenditures. In the usual order of things, men are in charge of all kinds of construction and equipment, especially if mechanization is involved, and they normally own the land on which any installation is built. This division of authority is a limitation on the likely take-up of toilets as a direct result of lessons on worms and germs primarily directed at women.

Thus it became clear that hygiene education, despite the very clear need for it, was not likely to be the key to the mass transformation of sanitary behaviour. Important though it was to understand faecal risks and dangers, that knowledge in itself was not enough to push large numbers of people to discard open-air practices and permanently adopt closed-cabin toilets. Not, at least, among adults. But among children – now that was another matter altogether.

To many in the industrialized world, the idea of there being nowhere for a child sent to school for the day to 'go' when the need comes upon him or her is extraordinary. Yet, shockingly, the vast majority of schools in rural areas of Asia, Africa and poorer parts of Latin America are built without toilets or access to water. In rural Nicaragua, where facilities do exist, they are often dilapidated, are not separated according to sex or age, are not suitable for small children to use, and do not ensure privacy – especially important for adolescent girls.[7] This is regrettably the norm all over the developing world – and even in some industrialized settings. In much of Africa and Asia, children are simply expected to hold themselves in – not helpful for their concentration. In break time, they must either run home – not possible if home is miles away – or find some piece of waste ground to squat on nearby. If there is a toilet, it may well be locked or reserved for the use of the teachers; or it may be so exposed to view that children are embarrassed to use it. In Nicaragua, 20 per cent of primary schools have satisfactory toilets and adequate water. In India, only one-sixth of 1 million rural primary schools have functioning sanitation facilities, and only 60 per cent have water in or near the school compound.[8]

To make matters worse, in most of these schools, not only in the two countries just mentioned but almost everywhere, the subject of hygiene and disease risk from faecal matter has until recently been widely ignored, in both educational and personal growth terms. Only when HIV and AIDS burst on the world did the question of what children were learning in school about self-protection from transmissible disease start to be taken seriously, and 'life skills' – including those connected to sanitation – begin to enter the curriculum.

At pre-school and throughout their educational careers, children in crowded classrooms and playgrounds are vulnerable to infections of all kinds, especially those connected to dirt. Time lost to sickness contracted at school because children were never taught to wash their hands and because the place they have to 'use' is filthy and full of pathogens damages not only health but educational prospects as well. Around 400 million school-aged children a year suffer from intestinal infections,[9] and hundreds of millions of school days are annually lost to dirt- and sanitation-related sickness;[10] in thousands of these cases, children actually die. Where youngsters are infested with parasites from contaminated faecal contact, their lives may not actually be threatened but they may have no energy and constantly doze off at their desks (Figure 5.1). One study in primary schools in Java, Indonesia, for example, showed that anaemia stemming from hookworm infestation could affect children's working memory and cause them difficulties in reasoning and reading comprehension.[11]

In many countries, the problem of lack of school facilities is connected to lack of resources, or perhaps more accurately in some settings, it stems from political and bureaucratic failure to commit the necessary resources to early childhood development and primary education and a lack of priority in social policy generally to children's wellbeing. Where there is an interest in children's education, other issues are likely to be considered – teachers' training, class size, text books, school equipment – before the minor matter of where children are to 'go' and how they are to keep their hands and bodies clean. Toilet blocks are not classrooms, so they do not come under the education budget; they are not considered as part of child healthcare concerns either, and they are left out of water and sewerage budgets too. As in so many contexts, sanitation in schools has fallen resoundingly into budgetary and departmental cracks.

When UNICEF and its international partner in school sanitation, the

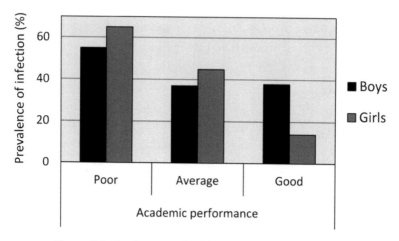

Figure 5.1 *The impact of schistosomiasis on schooling in Mali*

Source: IRC (2007) *Towards Effective Programming for WASH in Schools*, IRC, Delft; citing de Clercq et al (1998)
'The relationship between *Schistosoma haematobium* infection and school performance and attendance in
Bamako, Mali', *Annals of Tropical Medicine and Parasitology*, vol 92, no 8, pp851–858

International Water and Sanitation Centre (IRC) in The Netherlands, first began
to pay serious concern to toilet blocks and water supplies in rural schools, they
found that, on the ground, the issue was typically regarded as a matter of
construction.[12] But as a visit to schools in any deprived environment will testify,
if the facilities are not well built, and there is no commitment to maintenance,
hygiene information, soap and the doctrine of 'clean', they soon become unpleasant.[13] Even in industrialized countries, the poor condition of facilities is a
problem for many schoolchildren. Dilapidated toilet blocks in many urban areas
are connected to violent and bullying behaviour in schools, including assaults on
girls. Instead of helping children adopt good toilet practices and hygienic habits,
filthy facilities and threats of harassment or attack are bound to act as disincen-
tives to girls and younger children to use school toilet blocks. Teaching the virtues
of the enclosed toilet and hand-washing in schools is impossible in such circumstances. An entirely new, beefed-up approach was required.

In 1999, UNICEF and IRC launched a new partnership for school sanitation and hygiene education (SSHE), and the momentum and character of school
sanitation as promoted internationally began to be radically redirected. SSHE
was seen as a way to reclaim millions of lost school days, make schools more
attractive, and improve the cleanliness and personal habits of school-goers so as
to entrench them for the future. In addition, given the difficulties of selling
sanitation to the older generation, changing the behaviour of their school-going

children was seen as a potential way of changing behaviour in the whole society, if not immediately then over time. This would be attempted by bringing communities and schools, parents and teachers, closer together.

An important impetus converging with the new push for school sanitation was the simultaneous drive for 'universal basic education', one of the first big international targets and reiterated in the Millennium Development Goals. To improve school numbers required better schools: many children dropped out because their experience of school was so bad. 'Child-friendly schooling' became the new educational *cri de coeur*, with a revamp of the fabric and facilities of schools as well as changes in the curriculum and teaching methods. Attendance had to be made more attractive to children and parents, and teachers had to be encouraged to behave empathetically and inclusively towards all children, advantaged and disadvantaged alike. Too often, children who were not very clean or well turned-out suffered abuse and humiliation from teachers and peers, and left school because of this. Attitudes across the board needed to change.

Implicit in many reform packages was greater community involvement in schooling, with school management boards and village education committees enabling parents and local leaders to participate. Instead of reinforcing social exclusion, schools should be hubs from which ideas of inclusion could spread outwards into the community. Particularly significant was the realization that lack of toilet facilities was seriously inhibiting efforts to bring girls into school and keep them there. In many societies, girls might be withdrawn at puberty, or be repeatedly kept at home, not only because they were embarrassed about using the same toilet as the boys, but because they had no secluded place to change cloths or pads during menstruation.[14] Lack of separate toilets could expose them to 'talk' and loss of modesty, not to mention sexual taunting or actual attack. A girl in the Indian city of Pune described her problem thus:

> *The taps in the school all ran dry, and I needed to change [pads] every four to five hours for three or four days and hence I had to remain at home. One or two of my teachers were concerned about the gaps in my attendance and I was asked why I remained absent so often. Unfortunately, I did not have the courage to broach the subject, and I remained guiltily silent and accepted the blame.[15]*

The education of girls has a major impact on the whole of society. Take the state of Rajasthan in India. Here, female literacy is 32 per cent, compared to 71 per cent for males.[16] Girls suffer acute discrimination in this desert society, where poor care in early childhood means that Rajasthani girls have a lower survival rate than their peers almost anywhere in the country. The population of Rajasthan is sparsely settled and mostly tribal, still living in the fiercely independent way typical of peoples in remote and difficult terrain, whose cultural traditions have been honed through centuries of hardship. Most girls are married by the age of 15: in a typical village school there are some girls in their early teens with the tell-tale blaze of vermilion powder at their hair parting to show that they are formally already wed. When they reach adolescence, parents may no longer wish to let them participate in an environment where they mix with males and risk losing their purity and reputation – a fear augmented where there is no girls' seclusion or private space. Without education, girls become mothers in their teens, repeating the pattern of discrimination against girls from the earliest days of their daughters' lives. Retaining girls in school improves not only women's status and sense of self-worth, but the chances of their having smaller families, raising healthy children and establishing wider connections with society at large. Thus making schools girl-friendly, including by providing them with their own toilet and personal washing facilities, helps improve girls' education. This in turn helps to postpone marriage and pregnancy and gradually raises female status.

Obviously, some schools in any programme do better than others. A high performer has been the primary school at Durgapura, a village off the beaten track in Tonk, a poor district in Rajasthan in which a 'package' of health, water, sanitation, drought relief and educational interventions has been introduced in partnership with local NGOs. The school in Durgapura used to be situated next to a burning *ghat* – a place where bodies were cremated. The students had to use the *ghat* as their toilet. So when SSHE came along, the *ghat* was shifted, the compound cleared up, and toilet blocks for girls and boys constructed. Then the new school management committee produced extra funds to hire more teachers and build additional facilities. In 2000, 134 children were enrolled; by 2002, numbers had risen to 204, including a higher proportion of girls.

Every day, teams of sanitation scouts carry out regular duties: sweeping of school premises, gardening, garbage disposal, and cleaning of toilets and the handpump area. Once a month, a sanitation team goes around the village to talk to householders on the register of 80 families whose children attend the school. Households are checked out to see whether they are practising hand-washing, whether they keep all fingers out of the drinking water jar and whether they have installed a toilet. By late 2002, 27 per cent of households had installed a toilet, while 90 per cent were using a long-handled drinking-water ladle to keep drinking water clean. Within the school, posters adorned the classroom walls citing the virtues attached to cleanliness in all religions. Even though it takes time for families to change certain habits, and not every parent could be instantly persuaded to abandon open defecation, the school – under the leadership of a dynamic headmaster – was helping to inspire local pride in a clean living environment, and to establish habits which students will want to take forward into adult life.

This is just one of the districts in India where SSHE has been used as the entry point for the promotion of sanitation in the community. Since the original programme in 1998–2002, school sanitation has been incorporated into the government's 'Total Sanitation Campaign' and is now operational in 529 districts countrywide. In another Rajasthani district, Alwar, the improvement of school facilities has shown a marked connection to the enrolment rates of both sexes – but especially girls – and to school performance (Figure 5.2).[17] In Mysore District, one of the original intervention districts in Karnataka, the programme became so successful that international donors and state officials flocked to visit. In one block (sub-district), 110 out of 235 *gram panchayats* (local councils) applied for the programme – a clear indication of its popularity. The *gram panchayat* had to find 50 per cent of the costs, with UNICEF contributing the other 50 per cent, along with funding training, orientation and 'software' materials. Here, as in Rajasthan, student 'cabinets' were elected, with ministers for the environment, finance, culture and so on. They organized the cleaning of the school, planted and tended its garden, and saw that the 'facilities' were well managed. In some cases students were so enthusiastic that they contributed their pocket money to buy soap and brushes. The programme's popularity with local councils attracted official attention, and politicians began to request that the programme be extended to their constituencies. All of this was very gratifying to the programme promoters, who had initially fought inertia to get school sanitation off the ground.

However, there was a special twist to the story, illustrating that the process of scaling-up is never straightforward. Karnataka is a state with a rich agricultural output, and there is a high premium on land. A number of schools were established decades ago on pieces of land given for the purpose by a local landowner. Such land was usually regarded as waste, a dumping ground where animals might stray, where people might defecate or throw rubbish, and which no-one bothered to look after. The school buildings were typically huddled in a corner, from where they expanded into the rest of the plot. Today, with incomes and land prices soaring, many original 'donors' are trying to dispute the perpetuity of the grant and get these lands restored to their ownership. Many schools in Mysore district

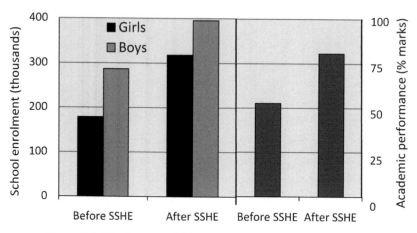

Figure 5.2 *The impact of SSHE in Alwar District, Rajasthan, India*

Source: Sumita Ganguly (2005) UNICEF SSHE Presentation, 2004–2005, New Delhi

are thus involved in litigation over their perimeter boundaries. Since the SSHE programme included money for the construction of boundary walls, nominally to protect the new water and sanitation facilities, schools and local councils saw this as an opportunity to consolidate their holdings and protect their area from its continuing use as a waste or defecation ground without having to pay all the costs for the construction of the boundary walls. These walls, however necessary to the school's wellbeing, were very expensive – much more so than SSHE software – and were not really anything to do with sanitation but with school buildings and security generally. What was making the programme so popular was the boundary walls, not the transformation of children's and parents' sanitary behaviour.[18]

Too often, when programmes with excellent track records such as SSHE are 'taken to scale', all the careful motivational activity – the development of sanitation scouts and ministerial cabinets, the rotas of student cleaners, the monthly community drive to encourage new 'adopters' – disappears. Instead, what happens is a bonanza of construction. Never mind the transformation of mindsets and sanitary mores, forget the effort to undermine the Great Distaste. On come the contractors, and we are back to COW – 'contractor-oriented work'. Now that school sanitation is being introduced on a wider scale into other parts of the country, sustaining the quality of the school sanitation inputs is a real programmatic headache.[19] Efforts have been made at the state level – special training for communications functionaries throughout the country, for example – to try and ensure the integrity of the programme as it expands. Prototype materials based on the pilot programmes were developed and distributed, including manuals, posters, games and other activities to strengthen hygiene education. But in the end, unless the school and community leaders are motivated and there is real commitment from local councils and education committees, all the careful planning and coordinating work at national and state levels may not fulfil expectations.

Using local NGOs to conduct familiarization workshops, impart skills and get schools seriously locked into SSHE is one way to make things happen. But where there is no effective NGO and things are in the sole hands of bureaucrats and contractors, numbers of constructions will always take precedence over behaviour change and sustainability. As assessment after assessment has emphasized, construction carried out without the software elements – hygiene education, school scouts, 'ministers' and their teams, life-skills instruction – leads to poorly maintained and dirty facilities, unhealthier schools, and little change in children's and their families' hygiene behaviour.

Although SSHE or WASH in schools has taken different forms in different countries, in key essentials the programmes which really work imitate each other. Well-designed facilities, separation of boys' from girls' facilities, and age-sensitive sizes of toilet are some of these essentials. Another is the element of 'participation' by students, teachers and community members. Participation is a buzzword in sanitation as in all areas of social development programming. It means getting people involved and helping to build their interest and therefore their responsiveness to any intervention. In Karnataka and Rajasthan, the school scouts and monthly sanitation drives are typical ingredients of sanitary participation.

Another has been the promotion of school health clubs, a feature of many successful programmes. In Kerala, south India, an NGO called SEUF (the Socio-Economic Unit Foundation) has prompted the establishment of 1230 health clubs in schools where they have constructed facilities.[20] SEUF has been imaginative in expanding the participation concept; for example it recently held a one-week student camp, in which groups of students observed the sanitary behaviour of local authority officials – including where they urinated and what they did with their rubbish – and wrote it up in a report. In a coastal neighbourhood, where fishing is the main livelihood and incomes are very low, SEUF also held a worm-infestation camp. Many local children were anaemic. When the health club participants tested their stools, and showed both the children and their mothers through the microscope the squirming livestock present in their faeces, the effect was electric. Children with parasites were treated at the local primary health centre, and demand for household toilets dramatically rose. SEUF has also pioneered another concept which is gradually catching on elsewhere: the installation of small incinerators in the girls' school toilet facilities to help them dispose of soiled menstrual cloths.

School health clubs were also a key feature of a project covering seven primary schools in Kisumu District in western Kenya. This crowded and fertile part of the country, bordering Lake Nyanza, has been notorious for epidemics of sanitation-related diseases, including diarrhoea, malaria and even occasionally cholera. Despite the high population density, few sanitary facilities existed until recently, and water drawn from open sources was rarely boiled before drinking. In 2001, Sustainable Aid in Africa (SANA), an NGO committed to community development through participatory methods, began promoting hygiene awareness within seven local schools. School health clubs involving staff, students and parents were formed. A one-week training course for the members was held, using locally developed toolkits and encouraging them to write their own songs, skits and health message posters. Boys were expected to share school-cleaning duties with girls. The clubs supervised the construction of gravity-fed water supply systems and VIP toilet blocks. To extend the programme into the wider community, SANA also trained ten 'village resource persons' in the area surrounding each school, whose task was to visit households and impart the same health and hygiene messages. These include injunctions to build and use a toilet, to wash hands after using it and before eating, to protect drinking water and food from contamination, and to keep the domestic environment clean.[21] Since the facilities were built, more girls have been staying on at school.

In Nicaragua, a different form of participation has been central to the 'Friendly and Healthy Schools Initiative' launched in 2001 under the umbrella of ENACAL (the state water company). The strategy is to work with associations of parents and with the school governing body to mobilize them around sanitation. When ENACAL's people begin the programme in a new locality, they hold meetings at the school with parents, teachers, members of the school council and students. A committee is then formed with representatives of all groups to carry out a 'community assessment' of existing attitudes and facilities concerning toilets, hand-washing, drinking water safety, refuse control, risks from animal faeces and wastewater drainage. Household visits are undertaken to enquire into male, female and young child behaviours. Local leaders are interviewed to understand their perspective on environmental cleanliness and local pride. Community meetings are conducted in which ENACAL social workers use participatory methodologies such as mapping, sketches, songs and games in which people vote by secret ballot on who are those most responsible for dirtying the community and what to do about it.

The process is designed so that the community can see what their environmental problems consist of, and that it is within their own power to reduce disease risks and make their homes and community spaces cleaner and nicer to live in. After this assessment is complete, the committee prepares an action plan, and the stage is set for the design and construction of new facilities, including at the school, under its supervision. 'At each stage of the process, the community identifies what to do, when, who will do it, how, and what resources they will use. If there are no resources, they work out ways of obtaining support from outside – maybe from an NGO, or the Ministry of Health, or by application for a loan,' says Elisena Medrano, ENACAL's programme director in Matagualpa. The construction of drainage channels they can usually manage for themselves. Installation of toilets may require special financial help. The committee may introduce regulations about where waste should be deposited, and about fencing so that animals cannot stray or get into people's houses. Arrangements may be made to collect vegetable refuse and feed it to local livestock. Domestic water filters and washstands are also promoted; these require soakpits or household wastewater drainage.

Thus sanitation in schools is closely integrated with environmental improvement in homes and in the community generally. Monitoring is included in the action plan, to see that what is planned actually gets done and to follow up reluctant householders. Reminders are issued and penalties imposed if obligations and responsibilities are ducked. The strength of the approach is that the sanitary idea is fully explored and implemented jointly by school and community, and if standards subsequently lapse, a determined councillor or school director will know how to get things mobilized again. The local municipality is always fully involved, and messages are reinforced through health centres and local media. Up to now, the programme has involved more than 200 schools and surrounding communities, reaching out to over 30,000 students plus their families and neighbours.[22]

Across the Atlantic, on the far western bulge of Africa in a landscape once 'discovered' by Portuguese explorers and still today dominated by baobab trees, mangrove swamps, sea-going pirogues and huge rivers disgorging into the ocean, is a very special school sanitation programme. In Casamance, southern Senegal, a long-running civil insurgency has disrupted regular life and inhibited development in every area. But an extraordinary effort based on Zinguinchor, the regional headquarters, has been made to keep the schools functioning and even enable them to be improved. Due to the state of poverty and the emergency, the World Food Programme (WFP) provides rations for school feeding for those in the most difficult and deprived areas. In 1999, UNICEF added a programme in the same schools called 'Building for Life'.[23] Its core feature is the construction of water-points and toilet blocks to reduce diarrhoeal sickness and worm infestation among the children, and generally demonstrate the virtues of safe and tidy toilet practice. The water supply also makes it easier to prepare and consume the daily school meal hygienically, and to wash up afterwards. But – as with all good SSHE interventions – the whole package is designed to bring about many other benefits.

First there is the usual need to encourage enrolment and retention of girls (in Senegal as a whole, 500,000 school-age children do not attend school, of whom 300,000 are female). And then, especially in this region of insecurity, there is a need to revitalize the schools and enhance educational attainment. The parent–teacher association is given a boost, and links between staff, students and the surrounding community are enhanced so that the school becomes an oasis at times of emergency or rebel disruption. Life-skills teaching for the students includes stress management and conflict prevention, environmental knowledge, and health and hygiene information. The water supply enables the schools to grow vegetables and constitutes an emergency standby for the community in times of drought. And in some schools, people from those households who have not installed their own toilet can use the school facilities out of school hours. Up to March 2007, 310 out of 435 schools with feeding programmes had been included, reaching 108,000 children.[24]

At Dar Salaam Pakau in Sedhiou District, the school is in a grove of vast baobab trees at the heart of the village. Although the director, Abjant Ndiaye, does not like the way the road leads straight through its grounds, this does mean that the school is right at the hub of village life. Ndiaye says that the main change brought about as a result of the programme is an expansion of school governance. There is now a school assembly with deputies from each class and ministers with portfolios: health, hygiene and sanitation, and apprenticeships, for example. Commissions are in place for the maintenance of the toilet blocks and for the water handpump. A woman from the village looks after the pump, and she and others have been developing the vegetable garden. There is also a commission to manage conflicts in the village. 'Since the school government was installed, there has been a much better sense of civic responsibility in the community and the atmosphere is much improved,' says Ndiaye. 'Once we came back from the weekend and found the toilet dirty, so we had a meeting in the village to change this.'

Although the primary targets are the children because they will hand on the ideas to their own children, Ndiaye believes that they transfer information effectively to their elders too: 'We organize the children to spread ideas in a democratic way. They wear special caps to give them self-respect. This has also changed their relations with their teachers, who are now willing to share some of their power.' The mothers in the village have also been prompted by the new fad for governance to start their own association. As well as supervising the school feeding programme, they monitor the way the children clean the school, and if the toilets are not as they should be, the commission for hygiene is summoned. When there is a festival, the school toilets are thrown open to all. The sense of high community morale, centred on the school, is palpable and convincing.

The design of the toilet blocks in Casamance is of special interest. In spite of the fact that this is an area with heavy rains in the wet season, the blocks are without roofs. A local NGO, Idée Casamance, which was set up in 1989 in an attempt to provide entrepreneurial skills to young men leaving school, has been responsible for their development. Since young people are usually asked to do the cleaning up at home, the original idea was to build on that and have them promote and construct toilets. But demand for household sanitation was low in the early days, and the 'Building for Life' programme has been their principal recent employer.

Idée's models have been based on the VIP, whose fly-reduction system depends on cabin darkness. But in the dark, it was found that boys aimed poorly. The walls were not tiled, and as urine is very aggressive, they soon began to smell horribly. First windows and light were introduced. Then it was realized that, if the blocks were built without roofs, during the wet season when the schools are closed, all the cubicles and pans would be thoroughly washed by the rains. The buildings also manage without doors since the student entering the block turns right or left to effect a discreet U-turn into one of the cubicles. This reduces problems with broken doors and hinges. Outside, there are washbasins with taps and soap. The whole construction is neat and compact. The entrepreneurs trained by Idée are now beginning to reap long-term rewards by obtaining orders for individual household bathrooms and toilets. Gradually, due to this programme and to a new effort to promote subsidized household sanitation by the Regional Sanitation Department in

Ziguinchor, opportunities for these builders of toilets, wastewater soakpits and washrooms are growing.

These examples, and many others from programmes in different settings, convey an encouraging picture of school sanitation improvement. But it is important to realize that many millions of schools have yet to be reached, and that the whole effort to install and upgrade facilities can fail if schools and the authorities to which they are answerable do not make an ongoing commitment – both in terms of constructing well-designed and easily maintainable facilities, and in terms of pursuing hygiene education within the school curriculum and culture. Alongside these efforts, others are required to promote similar approaches in day-care and early childhood learning centres with appropriately sized facilities, and to convey to mothers the particular need to use potties with their smaller children and dispose of their toddlers' faeces hygienically.

This section began with the idea that persuading children to take up new behaviour and habits in relation to sanitation and hygiene might be easier and more long-lasting than efforts directed at adults and that the school and its students could be a force for persuasion at home and in the community more broadly. This notion, that teaching schoolchildren to use toilets and practise hand-washing might be a short cut to bring about adult behavioural change, is still debated among experts. Where a programme is well designed, well integrated with the community and receives their support – as in the programmes in Rajasthan and Kerala in India, in Matgualpa, Nicaragua, and in Casamance, Senegal – the impact on the community at large is undeniable. But even where WASH in schools really is a practicable route to general sanitary transformation, the effort needs to be comprehensive and enduring. Results will be reinforced if sanitation and hygiene practices requiring the management of infants' faeces, toilets, showers, soakpits, drains and basins are promoted simultaneously to parents and householders, schools, markets and other public places in the community. In the examples explored, the school and its governance mechanisms linking it into the community have managed to act as a catalyst. In other cases, simultaneous programme efforts in schools and in communities reinforce one another. Systematic messages from several directions at once – teachers, health staff, priests, local councillors – cannot be bettered.

That has been the strategy of Association Miarintsoa, an NGO in the small market town of Antanifotsy in the Madagascan highlands. The local Association Miarintsoa team of social motivators and technicians has been promoting water and sanitation improvements in the community, schools and public places for the past year and a half. Up the road from their office, the head of a community of 24 modest households conducts a tour of well-kept toilets, and all agree that diarrhoeal disease among the children has substantially dropped since the facilities were introduced and paths became free of human dirt. During class break time, the director of one of three local primary schools involved in the programme gives an impromptu tour of his classrooms. At the far end of the playground, toilet doors are opening and closing and, without any awareness of the visitors' interest, boys and girls are washing their hands in the circular trough around the water pump. Do his students all use the facilities? Has the recent programme taken root?

> *They have all accepted. And they all take turns to sweep the classrooms and keep the toilets clean. Before, our schools had no facilities, there was nothing but the bush. Since the Association came to Antanifotsy, everyone in this place has been convinced. Here and at home, they all use these facilities now.*

Even in some of the poorest and most recalcitrant environments, the Great Distaste is finally under pressure. Without any doubt, the participation of schools is making an important contribution.

If schoolchildren have been a very important target of participation in hygiene transformation, women have been equally so. Women are responsible for everything to do with domestic management of water and children's toilet and hygiene practices at home. Where they have no tap of their own, they have to fetch all the water for the household, often in heavy containers over long distances, and are understandably careful about every drop. They are therefore gatekeepers for its use in hand-washing, bathing or personal grooming. Their own need for modesty and privacy, particularly in societies where women are traditionally confined to the home and a life of domesticity, also make them prime candidates for toilet demand creation. But when it comes to the addition of social mobilization to health education as a software tool, women have other attributes. They are good organizers and promoters. Once convinced of the virtues of the 'clean community' and equipped to tackle the Great Distaste, they are willing to visit neighbouring households and bring reluctant community members into line. Without this kind of social underpinning carried out by women, many of the sanitation drives of recent years would not have led to any real and permanent change in sanitary behaviour.

The techniques of social mobilization were systematically developed in the late 1980s, in an all-out effort to reach the target of 'universal immunization by 1990'. This worldwide drive for immunization, spearheaded by UNICEF, was in many essentials a throwback to the mass disease campaigns of the 1950s and 1960s. The idea was to energize the whole society behind the target of child immunization using all possible actors and modern communications channels. The strategy usually required there to be a National Day or Days, on which media frenzy would be guaranteed by a keynote Presidential 'event'. There would be campaign messages, public announcements on TV, banners, T-shirts, celebrity launches and platforms with garlanded dignitaries. The police, army, Red Cross, NGOs, religious leaders, politicians, corporate backers – everyone who might lend their name, weight or organizational resources to putting over the message – would be enlisted. Leading roles were always given to women's and children's organizations, including schools and sports and youth clubs. An example where social mobilization for sanitation was prominently used was Bangladesh in the early 1990s (see Chapter 3). Every year for several years, there was a National Sanitation Day, launched by the President, in which hundreds of NGOs organized rallies, marches and meetings around the country, and mass community commitments were made to toilet construction. For women usually confined to their homes, the organized rally or march – a form of public action well established in the Indian sub-continent – was a way of getting out of the house, respectably and in solidarity with others, for a positive community purpose.

Thus women were in the frame for all types of social mobilization connected to sanitation in which better health and family wellbeing were the principal messages. Although many social mobilization enthusiasts put a lot of stress on 'participation', however, they usually just meant 'joining in'. In most social mobilization exercises around National Days, women took part because they were exhorted to do so by those in positions of influence over them. Whether they had at the same time been 'empowered' to join in by taking their own decision to do so was not something that social mobilizers usually thought about. In traditional societies, women expect to obey those in authority over them, male and female. In most social mobilization campaigns, there was no effort to break this pattern, but rather to harness it. When the president, police, army and religious leaders are endorsing an action or message, the encouragement of independent-mindedness or a thought-through choice to do something is not part of the psychology: the idea is to build momentum and social endorsement behind a particular action. Social mobilization did not empower women to challenge decision-making norms in the home, even though this may be required to bring about such significant behavioural changes as enrolment of girls in school or construction and use of home toilets. Taking a child for immunization is an action of a different order. Unless people are independently convinced about a lifestyle change, they may well revert to old behavioural patterns when a mobilization exercise or campaign is over. Thus social mobilization on behalf of toilets and hygiene – as behind other lifestyle changes – has limitations. Nonetheless, it has helped to familiarize people with the idea, and speeded up sanitation spread.

In Myanmar, for example, National Sanitation Weeks were launched in 1998 to mobilize people to build their own sanitary toilets on a self-help basis.[25] In the first few years, the strategy was successful in reaching better-educated groups, and the rate of construction rose by 10 per cent annually in the towns and 5 per cent in the countryside. But those in poorer groups and without access to television and radio were almost untouched. Those in charge were also concerned that the achievements had more to do with people's aptitude for following political directives than with a process of building 'demand' on the basis of improved health and social wellbeing. They realized that if people did not appreciate the value of a sanitary toilet, they would not maintain it and the annual campaign outcome would not be sustained. The strategy of national weeks with strong political injunctions over a short period was therefore replaced with year-round social mobilization. Schools were encouraged to participate with drawing competitions and photo exhibitions on themes such as 'clean environment'. Partnerships were built with NGOs and community groups, small-scale producers of toilets and soap, and media organizations, including the Myanmar Motion Picture Organization. Building a sanitary revolution in Myanmar could not be achieved on the basis of national directives, however influential these might be. People had to understand the connections between lack of hygiene and diarrhoeal and other types of disease, and the role of sanitary toilets and hand-washing with soap in ridding themselves of these afflictions.

In southern Africa, the story of social mobilization took a rather different path. In the 1990s, a determined effort was made to introduce participation into community sanitation, hygiene and water management schemes via a training method known as PHAST: participatory hygiene and sanitation transformation. PHAST was developed and promoted by WHO and the UNDP/World Bank Water and Sanitation Program (WSP) as a methodology to be used with community groups, especially with women. PHAST deliberately set about 'empowering' people, not simply informing and educating them, or telling them what to do; it was supposed to provide the missing link between women's absorption of health education messages and real action to do something about them.[26] From 1990, the methodology was introduced into countries in eastern and southern Africa, on the basis of support from UNICEF and other international donors. The idea was that, at the end of the mobilization process, the group of participants would have taken on board sufficient new information and experience to be able to address their own problems with sanitation, hygiene and water protection. This was the counterpoint to top–down and supply-led approaches in water and sanitation, and PHAST's enthusiastic promoters bent over backwards not to prescribe things *for* participant groups or tell them what to do. The hope was that their new self-confidence, joint experience and desire for change would enable them to act independently.

In Zimbabwe, one of four countries to adopt PHAST, 800 environmental health technicians and 3800 health extension workers were trained in the years up to 1997, and during this time scores of PHAST workshops, with their emphasis on visual aids, creativity and releasing latent energies, were conducted up and down the country. In the following years, however, although the concept had become well known, the number of community-led programmes that took off proved disappointing. Unless individuals are exceptional, it takes more than a short training course to become lastingly 'empowered'. Many district staff failed to incorporate the PHAST activities into their working style and programmes – they found them too labour-intensive and time-consuming, and too reliant on extrovert behaviour and personalities.[27] Donors tend to look for instant results, and it was in the nature of an approach focusing on psychological processes that these were not forthcoming. PHAST was a pioneering participation methodology, rather than an applied programme with tangible outcomes. By the end of the 1990s, belief in participation as a motivating force was undimmed; but after a decade of experience, those originally behind PHAST came to the conclusion it was not achieving enough, fast enough, and donors began to drop away.[28]

PHAST showed that empowering women to participate was not sufficient on its own. Some ongoing programme or forum that they could participate in – without being too prescriptive – was also required. For a participatory approach to work, a balance is needed between evoking energies and channelling those energies, especially in societies where women are unused to community leadership roles. A new attempt to capitalize on the existing training was begun in Zimbabwe in 1995 – with the difference that mobilization would be linked to actions with measurable outcomes; groups would not just be left to do their own thing, but would undertake specific tasks. The key to the approach was regular participation in a 'club'. In countries where British missionaries and philanthropic societies had made their way during the colonial era, there was a long experience of women's clubs, modelled on the archetypes of the Mothers' Union and Women's Institute. Now, the concept was slightly adapted to take the form of 'community health clubs', also promoted through the extension health worker structure, using PHAST materials and methods. Although membership in these clubs was not exclusively female, women constituted the overwhelming majority of active participants.

Every fortnight, members met for an evening of socializing and listened to a local speaker – usually the environmental health technician – on a health-related topic. Each woman had a membership card listing the topics. At every session a piece of homework was set for next time – for example providing a water storage jar with its own cover – and points were given for achievement. After around a year of meetings, club leadership usually gained strength, and at this stage more significant changes at home – toilets, washstands – would be introduced. An NGO was set up to support the development of the clubs, training the speakers and leaders and equipping them with visual aids; members received T-shirts and attendance certificates. By 2000, over 500 clubs had been started. In one district, Makoni, 2400 toilets had been built within two years among 11,450 club members.[29] Where members did not build toilets, they used the cat method, or 'faecal burial', a method of sanitation that was new to them. The programme managed to build a 'culture of health': everyone built a drying rack, or covered their water or constructed a VIP, because it had become a community norm and you might become a social pariah if you did not.[30] Thus mobilization with an added structure of activity had worked where 'participation' alone had not done the trick.

Community endorsement of new behaviours is now widely seen as an important way to promote sanitation to new users. As many examples have suggested, if the behavioural norms of the whole community can be tackled and changed at one time, the results are much more likely to be lastingly effective. However, for this to work there may need to be special characteristics operating in the community, or special influences operating from outside it. Factors internal to the community include high population density: in most rural environments where community mobilization has worked, people live in a closely settled pattern, whether they live in an arid area or a highly fertile one where land is prized. This is often an area where vegetative cover has been reduced by population pressure. Such areas are similar to urban locations, where demand for sanitation barely needs to be created – it already exists. Another factor is the position of women. If women exert personal influence over their husbands or social influence as a group, and if there is a relatively high enrolment of girls in school, the chances of mobilization are higher. Among other external influences are political leadership, the nature and degree of civil society organization, and the penetration of the cash economy. In West Lombok, Indonesia, and West Bengal, strong political commitment, powerful leadership and the solidity of local institutions has been critical, as was the presence of spare cash so that home improvements were affordable.

A final example of social mobilization for successful sanitation, from Ethiopia, illustrates a combination of these factors. In the south-west of the country is a region called the Southern Nations, Nationalities and Peoples' Regional State (normally shortened to Southern Region); it contains 20 per cent of the country's population – 15 million people – in 10 per cent of its land area. As its name suggests, this region is home to a wide diversity of cultures and ethnic groups, some living in very densely settled areas. In early 2003, the proportion of households with pit toilets was 13 per cent; within a year the rate had gone up to 50 per cent and within two years to 77 per cent.[31] Here is a case where critical mass for adoption of sanitation was achieved over a very short time-period, largely on the basis of stepping up the use of an existing primary healthcare (PHC) network with a strong presence in the community.

The impetus for a concerted campaign for better health came from the Regional Health Bureau, under the leadership of Dr Shiferaw Teklemariam. Prior to 2004, Dr Shiferaw saw the prevailing health situation as a 'leaking bucket': 'The rural people got sick, they were treated and left [the health centre], then they got sick again, and were treated again ... and the cycle continued. They spent most of their cash income on healthcare.' Volunteer women health promoters were trained to work at the outer edge of the existing PHC service delivery system, alongside health extension workers and backed by local community leaders. Dr Shiferaw decided to use this structure to focus on six high-impact interventions, including mother and child healthcare, family planning, immunization, and – toilets. He gained support from the regional political apparatus with the ratification of a Regional Public Health Proclamation and coined the slogan 'Sanitation is everyone's problem and everyone's responsibility'.

The PHC personnel were prepared and trained at all levels to support the six-interventions campaign. Staff signed up to a contractual agreement to deliver on the six interventions and reach certain targets. Then the work of mobilization began. Due to the crowded landscape and acute deforestation, there was known to be a latent demand for sanitation, especially among women. They complained about how often they encountered faeces in the banana plantations and in the fields where they gathered fodder for cattle. They also complained of the smell, and of the embarrassment of seeing people defecate in the open. But until this point, there had been no priority attached by health officials to the problem of promiscuous defecation, nor any solution offered for the problem to be addressed. Now women were invited to do something about this habit and the mess it produced, and encourage others to follow suit. The volunteer health workers, on whom fell the critical outreach work, first installed toilets in their own houses and went on to promote them throughout their neighbourhoods. Agreements were also made with local leaders to reach set targets at the community level.

The type of toilet they promoted was absolutely elementary and basic: a platform with a hole; a pit underneath, lined if the soil was unstable; a cabin on top with walls and a roof. No subsidy was given: the costs were no more than a few dollars for the platform (with a superstructure built of natural materials), which most people could afford. Most of these toilets really are basic latrines: not many are 'improved'. However, the idea of managing your own excretion, containing it near your dwelling and not putting it out into the environment to inflict health risks on others is a significant social change. Those who are infirm or disabled and not able to dig the pit receive help. There are regulations: the toilet must be behind the house and positioned so that the wind does not blow towards other buildings; it must be a certain distance from any water source; there should be a ditch to protect the pit from flooding in the rains; and there is supposed to be a hand-washing facility. The regulations are not always observed, but at least a step has been made towards mass household change: a household without a basic pit facility is no longer quite the thing.

Ethiopia has a highly organized local government structure, with a clear line of command and strong tradition of mass participation. A hierarchy of responsibility stretches down through the regional, zonal, district and community healthcare structure and administration, right into the household, and the signed agreements make clear what each level is supposed to do. As in Myanmar, West Bengal and West Lombok, these structures and their ability to respond to dynamic leadership have proved very important in sanitary advance. Effective organization, incentive schemes (prizes), motivation and teamwork have all contributed to the pit-toilet epidemic in this region of the country. Nonetheless, these are still early days. Most of the toilets are very basic and their cabins are made of temporary materials. If they are not to suffer the problems faced in other settings – overflowing pits, structural collapses during the rainy season, a build-up of smells and filth – momentum will have to be maintained. People need to be encouraged to move up the 'sanitation ladder' and convert their new slab toilets into 'improved' facilities: perhaps add vent-pipes, which are currently uncommon in Ethiopia; maybe add hole covers and fly-reduction systems; or possibly introduce eco-variations. This requires developing a range of technological options and a sanitation 'market'. At present, the production side has yet to be adequately addressed. There is also little sign yet of improved hygiene and hand-washing: new components will definitely need to be added.[32]

Since Dr Shiferaw – since promoted to Deputy State Minister for Health – and his PHC teams managed to achieve such a phenomenal rate of rural toiletization, every international donor has beaten a path to his door. New phases of activity are under consideration, with international support for software, training, technology development, the development of a sanitation policy, a regulatory framework and all the rest of the institutional structure necessary to move ahead. At this stage, subsidies for household sanitation, except for the very poorest and those with infirmities, are not in the plan. Dr Shiferaw believes that the explanation for what has happened is that a cultural revolution has taken place. 'Sanitation is not something you give away as a commodity,' he declares. But will this cultural revolution endure beyond its initial phase? And what will happen when household expenditures on bathroom 'improvements' and more permanent structures are required?

In one particular, his views are not echoed by leading purveyors of either the 19th-century sanitary revolution or that of today. Sanitation – or rather the part of it comprised by a toilet – is a commodity. That is precisely what a toilet is. And it is to the consumer dynamics of toilet demand that we now turn.

The circumstances in Ethiopia's Southern Region showed that there might be rural demand for sanitation but that this demand was hidden. While demand for water supplies is often strong and vocal, even to the point of being a vote-catcher for politicians, demand for sanitation is usually inaudible. Unless social researchers go around with their 'knowledge, attitude and practice' questionnaires, and put the question directly and privately to householders, the taboos surrounding excreta mean that latent demand has no channel for expression. On top of this, lack of knowledge of the commodity which could improve matters inhibits any such expression. It takes familiarity with a new kind of product or service to want one. In the 19th century, the spread of the flush WC was initially a response to consumer demand. Its success as a consumer item helped to *precipitate* the sanitary crisis in rapidly urbanizing Britain; it was part of the problem, not the response. Sewerage, in the search for an effective mass system of disposal of WC outputs, was the solution on which civil engineers alighted. The toilet in its private cubicle or bathroom met a personal need; the waste disposal system was something else altogether, a public as well as a private good. Surely, there is a lesson here.

In the case of on-site pit sanitation, the questions of personal convenience for a private act and of excreta disposal for communal benefit are conflated. The toilet appeals to the consumer as a personal home improvement; but in the public health perspective the important role of the pit is to confine excreta and remove it from public spaces, not to meet a consumer need. Only if everyone in the locality digs a pit and uses it consistently is the public good aspect fulfilled. Hence the recent preoccupation with 'total sanitation' – meaning that everyone has access to a facility, appreciates its health protection value and uses it in place of open defecation – in many programmes. To arrive at total sanitation in a non-engineered environment requires everyone to construct an individual or shared facility. So the behavioural change required is not just to use a toilet as a personal lifestyle choice, but to participate in community responsibility for human waste disposal – a far more difficult objective, and one never sought in the sewered and regulated living environment of the industrialized world. However, the first issue to address – since it is the first stumbling block – is how to elicit and respond to demand for the private item. Without keen 'adopters', no programme promoting the toilet habit could progress. If private consumer demand could be generated, maybe the public aspect could be sorted out afterwards – just as it was in the industrialized world.

In the early 1990s, a researcher called Marion Jenkins set out to find out more about the dynamics of household demand for toilets in a rural location in Africa. The extensive study she undertook in Zou Department, Benin, subsequently became influential in paving the way for the 'social marketing' approach to sanitation spread. What Jenkins wanted to work out was why some households in a particular area had decided to abandon 'open defecation' – the prevailing local practice – and install a pit toilet at home, and why most others had not.[33] Those households who had installed toilets had done so entirely of their own volition. Messages about the virtues of sanitation had been promoted by local health clinics and community development programmes for many years, but there was no actual delivery programme. So the information about whatever demand existed would not be affected by availability of subsidized toilets or any other programmatic anomaly or distortion to natural product take-up.

Benin is a relatively poor West African country, and most of the rural population work in semi-subsistence agriculture. However, in Zou Department, the home of the Fon people and the headquarters of the voodoo religion, there are also skilled artisans and entrepreneurs. Informal employment, mainly in cottage industries, is based on semi-urbanized population centres. When Jenkins began to map the installation of household facilities in villages across Zou, she discovered that the key factor in the pattern of demand was proximity to the local town of Abomey-Bohicon. Within 3 kilometres, demand was 38 per cent, but outside this distance it began to drop rapidly until, at a range of 15 kilometres, it was negligible. Those householders who had gone in for sanitation had done so after being introduced to this improvement at the houses of others. It had taken many years after the first toilet in the area was installed – in 1954 – for other householders to follow suit. But after the early 1970s, the process had speeded up. The spatial and temporal pattern was very similar to the classic pattern of adoption of any innovation – electric light, television, refrigerator, washing machine – among consumers. If you lived close to others with a pit, slab and cabin, you were more likely to install one yourself.

Interviews with the heads of toileted households revealed that the decisive persuader was prestige. In the past, bodily functions were deliberately performed away from the home. But the modern way of doing things, as learned by those familiar with city ways, was to undertake this personal requirement within the confines of the household. This lifestyle behaviour, and the installation of the necessary facility, identified toileted homeowners as up-to-date and enabled them to entertain important visitors. Without such a device, no-one of status, especially if he aspired to mingle with chiefs and princes of the Fon royal house, would be able to hold his head up. Possession of the asset was therefore partly symbolic: it displayed connections with the wider world and indicated experience of ideas and attitudes encountered elsewhere. The installation of a toilet in the home also improved the value of the property in the eyes of other clan members by making it 'complete', and conferred a legacy on descendants. These values were important, especially among men, in whose hands control was vested over major household expenditure and infrastructural change. Women and the elderly or sick tended to rate more highly the wellbeing attributes, the convenience and comfort, and protection from the hazards of 'going out'. Where living space was becoming more constrained – also more common nearer to the town – and it was increasingly difficult to find 'good' defecation sites close to home (clean, private, safe and socially appropriate), this was an additional incentive.

Men therefore led the way as far as toilet adoption in Zou was concerned, much as the socially aspirant bourgeoisie of the British Victorian world would have wanted the latest bathroom equipment in their newly furbished, 'look-how-successful-I-am', homes.

This study helped to reposition household sanitation in the framework of better homes, and to put men and their decisive influence over home improvement, especially improvement involving cash expenditure, into the picture. If prestige and being up-to-date was an important driver for sanitary demand, then sanitation should be marketed on that basis. Jenkins thus put forward the view that demand for toilets could be built up more quickly by a promotional campaign which crafted its messages around the benefits which existing users most appreciated, and that wellbeing and privacy for women was one of these but not by any means the only one. She also asserted that it would be wise to start in communities nearer the town, whose members were both more familiar with toilets and had more access to cash. This approach virtually discarded health education as a promotional driver, looking instead to marketing techniques to match toilet products with incipient consumers. A subsequent marketing campaign for household sanitation in Benin used the slogans: 'A beautiful latrine: Privacy guaranteed for all the family', 'A latrine is better when you have visitors', and 'With a latrine, no risk of snakebite'.

Aesthetics were also important. Important problems uncovered by the study were the lack of durability of existing models and the bad smell. This is one of the major difficulties with many low-cost schemes whose primary driver is public health: the unpleasantness of keeping shit in your home or compound if the equipment involved does not remove its stink or visible presence does not seem to occur to those who don't encounter pit toilets in their own living environment. When the whole idea is unfamiliar, and, in proportion to household income, significant costs are involved, it is vitally important that the toilet is congenial to use, and remains congenial over time with minimum effort on the part of users. In Benin, the inventory of designs that people had installed showed a wide variation in styles and costs, and much personalization – just as with bathroom fittings in the standard industrialized home. Jenkins concluded that if there was to be any kind of mass sanitation uptake, there needed to be public investment in technologies for all, and the development of a market with a range of low-cost and higher-cost toilets, matching the wallets and tastes of all kinds of potential consumers.

Today's toilet missionaries have become enthused by the 'social marketing' approach to sanitation spread. They are starting to borrow ideas from

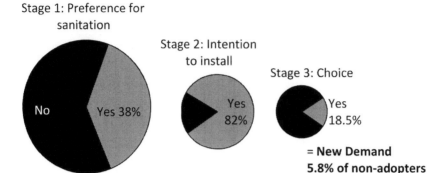

Figure 5.3 *New sanitation demand and adoption stages*

A study in Ghana set out to quantify demand for toilets at different stages of adoption. Many people in principle wanted to have such a facility. But that is not the same as actually planning to do so, or of taking action to install one. Only a small proportion of those in favour of household sanitation reached this stage, indicating that more needed to be done to make the process easier for the rest.

Source: Marion W. Jenkins and Beth Scott (2007) 'Behavioural indicators of household decision-making and demand for sanitation and potential gains from social marketing in Ghana', *Social Science and Medicine*, vol 64, no 12, pp2427–2442

commercial marketing gurus, examining what it takes to make someone who thinks that having a toilet might be 'a good idea' to take the step of finding out how to get one; and then what it takes to reach the final stage of purchase and installation. A recent study in Ghana which looked at demand in this stage-by-stage, carefully nuanced way found that, among those without a toilet, 38 per cent would – in principle – like to have one in preference to their current place of defecation. This was mainly because it would be more convenient, especially for elderly or sick members of the household, and it would be less squalid and dirty. Of these, over 80 per cent expressed an intention to install a facility; but only just under one-fifth of these respondents had made the decision to progress to the finishing line and build a toilet within the next year (Figure 5.3).[34]

The small proportion of those committed to actual installation was instructive. Enquiry revealed that it often had to do with expense and other practical considerations such as a lack of competent toilet-builders in the vicinity. If it is not easy to satisfy a new consumer desire, the desire may evaporate. The study suggests that, once sanitary promoters begin to look at what is preventing people in a given setting from moving from 'preference' for a toilet, through 'intention' to 'choice', and find ways of removing those obstacles, take-up can move faster. Bjorn Brandberg, the inventor of the sanplat, has also cottoned onto something very important with his Saniplast Privé (see Chapter 4): consumer appeal.

Recently, he has come to believe that, however highly you polish a concrete sanplat, it will never be so appealing – nor so easy to clean – as a nice, shiny, coloured plastic version.[35] Trendy plastic sanplats, purchasable in the market like the jerry-can or huge enamel basin, may in time become the equivalent to the corrugated 'tin' or *mabati* roof so beloved of status-conscious rural African housewives. The emphasis on housing improvement rather than health aid is critical.

In the end, the take-up of sanitation by householders is going to be a consumer-driven phenomenon. Without consumer interest on a reasonably comprehensive scale, community benefits and measurable health impacts are not likely to materialize. 'Software' for sanitation has not been an easy nut to crack, and where consumer wants and desires have been ignored, it has not been cracked at all. Hygiene education remains vital, especially in schools, primary healthcare centres and life-skills curricula for young people; it is not the case that health education messages make no impact, but it is decidedly true that they may not be a sufficient motivation to install a sanitary toilet. Social mobilization in health campaigns or sanitation drives also has a useful role to play, and clubs and societies fostering healthy living can, in a receptive environment, be even more effective. In other settings, strong leadership and community sanctions, backed by incentives – prizes and awards – may make a significant difference for a limited length of time. But only a drive to reveal, create, articulate and satisfy consumer demand is going to make possible the necessary sanitary transformation and entrench it over the longer term.

In order to do that, a market is required: a system of production, supply, marketing, advertising and consumption that puts goods out into the highways and byways that people want to buy and, having bought them, use and replace when they are worn out or when a better one comes along. A market economy around on-site sanitation for low-income consumers is what, actually, the sanitation revolutionaries of the 1970s set out to inspire. So why has it taken so long and what has to happen to move things faster in this direction? The pit-toilet economy – its investors, employers, workers and aspirational consumers – is the subject of the next chapter.

6

Shitty Livelihoods,
or What?

Previous page: Employment in the new sanitation consumer economy is one of the benefits of the spread of sanitation in West Bengal, India. At sanitary production centres, women are mostly employed to polish the pans, sometimes with younger members of the family, while men mainly cast pans and slabs.

Source: UNICEF Kolkata

In the days when public health arguments and supply-led approaches dominated efforts to sanitize poor communities, the economic benefits of investments were presented purely in terms of savings to the health budget, and gains to economic production in terms of person working hours and days. WHO has recently conducted an extensive exercise to calculate these benefits: for example, it has calculated that a US$1 investment in water and sanitation services yields between US$3 and US$34, depending on where it is spent, and that reaching the Millennium Development Goals (MDGs) for water and sanitation would save the world US$7.3 billion in health-related expenditure every year and US$750 million annually in the value of adult working days (Figure 6.1).[1] But equally, if not more, important gains of an economic kind have recently begun to attract attention. These relate to the sanitary consumer economy: the growth of entrepreneurship, manufacture and employment to respond to the demands of people who are set to become proper toilet users. Economically active sanitary personnel extend way beyond public health engineers and sanitation promoters to include those who build facilities, service them, and market toilets and waste disposal to potential customers.

No examination of the marketplace dimensions of consumer-driven sanitation should overlook the continuing existence of the 'sweeper', a euphemistic descriptor for a person whose job involves all kinds of rubbish removal, including shit. Night-soil collectors and manual diggers of raw faeces from pits continue to operate in a surprising number of societies, including in China, West Africa, Tanzania, Kenya, southern Africa, Bangladesh and Pakistan. But the most

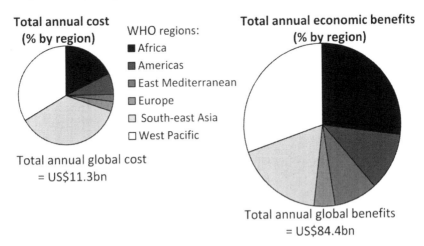

Figure 6.1 *Costs and benefits of reaching the MDGs*

Source: Guy Hutton and Laurence Haller (2004) *Evaluation of the Costs and Benefits of Water and Sanitation Improvements at the Global Level*, WHO, Geneva

obvious place to look for a person whose livelihood depends on the disposal of human waste is at an Indian train station. Bent double, down on the tracks off the end of the platform, a woman sweeps together human faeces into small piles. Using two flat pieces of tin, she scoops up each pile and puts it in a bamboo basket. She then puts this on her head, hoping that no liquid will fall in to dilute the dirt and send stinking drips between the bamboo weave onto her hair and down her face. Once the basket is full, she takes it to the designated place where its contents will be picked up by a (male) sanitation worker with a tractor. This woman is still plying her occupation today, in the 21st century, even as business-men with briefcases and mobile phones pace the station platform.[2]

Despite years of activism against the degradation of manual scavenging – over a century, if Gandhi's efforts are included – as well as many decades of activity by more recent campaigners, and despite a 1993 national law banning this work, around 790,000 people still live by personally handling human shit in India alone.[3] They work with bare hands, have no protective clothing, and use the most rudimentary tools. The broom and bucket may be handed to a bride by her husband's family on her marriage: 'At home I had never done this clean-ing job, but here I knew I had no choice,' said a woman sweeper interviewed in a recent survey in Andhra Pradesh.[4] The activist organization Safai Karamchai Andolan (SKA) puts their numbers much higher than the official estimate, at up to 1.3 million.[5] Yet several Indian states do not admit that this type of work still exists. Whichever figure is correct, the trend recently has been an increase – not the decline that one would expect to accompany burgeoning Indian prosperity and showpiece urban growth.[6]

Here is a traditional labour market associated with excreta removal; but civilized humanity demands that this workforce be put out of business to end the degradation and discrimination they suffer. In most parts of the world where such work persists, it is dying out. This is also the case in India, in the sense that dry facilities are gradually being replaced and sweepers becoming less visible; but from the point of view of organizations campaigning on their behalf, far too slowly. An irony of Indian sanitary progress is that it requires that the largest cadre of informal sanitation workers in the world be made redundant. This is because until that happens, there is no possibility of livelihoods associated with excreta being elevated to a position of social respectability or commercial success. For the creation of mass consumer demand for low-cost toilets, that will be essential. Among elite and middle-class Indians who live in an industrialized environment in which waste disposal systems are modern and upgraded, the issue does not arise or is confined to marginal moments (such as being 'caught short' in the street). But among the more modestly situated households in pre-modern environments, and in old-fashioned workplaces and other public or institutional spaces, the need for decent facilities to deal with the disposal of bodily emissions continues to be circumvented by the ubiquity of a cadre of people condemned to dirt removal as their function in life. There remains an unspoken assumption in many parts of the country that it doesn't matter where you put your shit or filthy debris (including cloths stained with faeces or menses), because someone exists whose job it is to take them away.[7]

The SKA and other Indian voices who excoriate the work of manual excreta collection focus on its intrinsic caste associations. In traditional Hindu society, caste and occupation were affiliated in a complex set of arrangements governing patronage, livelihoods, and systems of service and exchange. In the times of the Hindu and Moghul empires, there was no need for sweeper or scavenger 'employment'. But during the process of urbanization under British rule, the caste designation evolved into a 'job'. Large numbers of agricultural workers, excluded from the land as a result of colonial policies, were brought in as migrants to do this work, a process similar to that caused by globalization in labour markets today. This work became systematized all over the sub-continent, including in what are now Pakistan and Bangladesh. The army, railways, munici-palities, courts and industries established official scavenger posts, filled by members of the appropriate 'scheduled' or *dalit* castes, to which those handling carcasses and debris belong. So strongly has the social hierarchy continued to be influenced by ideas of purity and pollution that the customary degradation of such people and their 'untouchability' has never died out. Sixty years after special provisions for *dalits* were laid down in the Indian constitution, this prejudice persists. Its counterpart, a fatalistic acceptance of social servility by those who deal with society's detritus, allows the perpetuation of a system of faeces removal in which those who perform work which ought long ago to have been 'improved' into obsolescence may still be treated as if they were synonymous with the filth they handle. At school, their children are routinely ostracized and forced to sit apart.

Scavengers are still employed by municipalities, mining corporations and transport authorities to remove excreta from what are known as 'dry latrines': places where human wastes are left openly on the floor of an enclave built for excretory use, often on a concrete or stone slab, sometimes on a dirt floor, or dumped in a shallow gutter where excrement does not dry out and is much more difficult to remove. These 'dry latrines' have no dug pits; they have nothing to do with VIPs, composting or 'conservancy' and cannot be described as 'toilets' at all. They are places in towns and crowded spaces equivalent to the designated areas away from their houses that people in the countryside go to under cover of dark. However, unlike in the countryside, there is no natural process for waste neutralization or absorption. There are also many other built-up urbanized places that people resort to, including roadsides, railway tracks, derelict ground, building sites, alleyways and open gulleys, from which human filth and other refuse has to be removed. Thus the continuing existence of the sweeper category or caste is the reverse side of the continuation of 'open defecation': each of these practices props up the other.

They are also employed to empty private and public pits, just like the night-soil men who used to cart away the contents of foul-smelling cesspools in pre-19th-century European homes, or the household buckets of 19th- and 20th-century 'dry conservancy' systems. In the old quarters of certain cities such as Hyderabad and Patna, and in many smaller towns and built-up villages, the use of sweepers continues.[8] In urban Pakistan, they are known as *Churha*, the name of a Hindu caste linked to polluting work. Despite mass conversion to Christianity to rid themselves of the association, they still suffer profound discrimination, and while employment by the public services used to mean that their jobs and economic status were secure, nowadays, with privatization and the contractor culture, even that modicum of positive advantage may be removed.[9] Public divestment of services, along with failure to modernize rubbish collection and waste disposal management effectively, and factors such as declining rural incomes and exclusion from the land, explain the rise in manual scavengers among the casual workforce. Institutional apathy towards this caste-sanctioned method of human exploitation, and civil society indifference towards the plight of its workers, keeps the system going.

At the time that colonial occupiers built their municipal infrastructure and residential cantonments, sewerage as a city-wide system of human waste removal was still barely off the drawing board in Paris and London. The problems of environmental pollution caused by voluminous flows of raw excreta into waterways had also yet to be resolved, even in the more beneficent meteorological and hydrogeological environment of Europe. Many critics of human cartage and the 'dry latrine' look to flushing toilets and sewerage to provide the solution to scavenging and pit-emptying. But this is not a practicable alternative, either from the point of view of cost or from that of water availability. All the efforts of post-Gandhian warriors, *dalit* activists, socially minded NGOs and today's 'Total Sanitation Campaign' have not galvanized sufficient investment in other approaches. Ishwarbhai Patel, a famous toilet protagonist and opponent of sweeping based in Gandhi's home town of Ahmedabad in Gujarat, blames local municipal authorities for the perpetuation of manual scavenging, because they have failed to do away with the foul grottoes that substitute for decent public facilities. In 1991, a national planning commission task force declared that 'dry latrines' should be universally scrapped by 1995, but around 920,000 still existed countrywide in 2003. Another guilty party is Indian Railways. Over 30,000 of its coaches have toilets which discharge straight onto the track, and many station platforms have no washable concrete aprons. But no railway minister has yet provided funds in the railway budget to convert carriages and platforms.[10]

Within the movements active on behalf of scavengers over the past half-century, there has been a range of opinion about whether they should have their status, pay and conditions of work upgraded, or whether their employment in excreta-handling should definitively and absolutely cease. The Sulabh International *shauchalaya* movement, initiated in 1970 by Bindeshwar Pathak, defied tradition and prejudice by building well-managed and hygienic pay-as-you-go public facilities in cities all over the country. The toilet Pathak developed as the basis for his enterprise and to avoid anyone having to handle wet excreta (as described in Chapter 4) was the twin-pit pour flush; this allows the faecal matter to compost for 18 months, becoming a safe and inoffensive crumbly material before being removed. Sulabh toilet blocks, built in collaboration with state governments and municipal authorities, are now common all over India, near bus stands, markets, ports, parks and railway stations, and are used by upwards of 10 million people every day.[11] Pathak has potentially made obsolete the continuation of the 'dry' public facility; but so far he has not been able to break the mould of caste-associated occupations. His version of scavenger liberation is an upgrading of their jobs, including no cartage by head or other means. They are still employed to clean the toilets and empty pits, but other castes usually supply the manager-caretakers of the blocks.[12]

Bezwada Wilson, the convenor of the SKA, today's most active campaigner on behalf of the *safai karamcharis*, is uncompromising in his stand on total abolition. Other voices, notably Gandhi's, were in favour of regarding sweepers with ennobling admiration, treating them as 'children of god' (*harijans*), or proposing the amelioration of their working conditions – gloves, wheelbarrows, plastic buckets. Wilson repudiates this idea as a back-door method of perpetuating the practice and its accompanying attitudes. Although this position has not always been welcomed by all those who are employed as sweepers, fearful that their livelihoods will vanish, the SKA insists that all manual scavenging must end. In

2004 it undertook a campaign of physical demolition of 'dry' facilities in Andhra Pradesh, which succeeded in hastening the application there of the 1993 Act. But according to a survey conducted by the magazine *Frontline* in 2006, there are many states where enforcement languishes. Even on the outskirts of Delhi, scavengers are still summoned by officials or private individuals and expected, for the most meagre of tips, to empty people's pits or remove their filth from public spaces. Until this expectation of sanitary servility is renounced through-out society, it is difficult to picture significant public or private investment in the development of lower-grade, but improved and properly paid, sanitary occupa-tions of the kind that do not carry such profound stigma in other societies.

The taboos associated with the handling of shit in India demonstrate an extreme case of interactive connection between socio-cultural alienation from faeces and the difficulty of developing a new economy around low-cost sanita-tion. In the sanitarily improved Indian future, there will still be household pits, both single and twin, with raw faeces to empty in crowded residential areas. In high water-table areas, there may gradually be take-up of above-ground, double-vault, urine-diverting and non-urine-diverting toilets. In desert areas, VIPs may yet be accepted. But in all designs of on-site facility, unless the pit is very large and deep, emptying at some stage is required. Although composted matter from the eco-toilet or sealed-off fallow pit may be safe and inoffensive, someone still has to shovel it out. Someone also has to clean the toilet: this is inescapable even with the most pressurized flush and the shiniest porcelain in the world. At home the task will fall to women and their helpers. In other settings, it is difficult to picture caste Hindus taking on jobs keeping public toilets clean, even in such immaculately 'improved' environments as airports and five-star hotels. Meanwhile, if proper public and institutional facilities are constructed, and semi-mechanized pit-emptying services introduced, it will be difficult to maintain that ex-sweepers should be dissuaded from applying for upgraded sanitary workforce jobs.

During the protracted transition away from a caste-based hierarchy predi-cated on sanitary work at the lowest possible social rung, there will have to be some compromise between upgrading and abolition of livelihoods in which management of human excreta or toilets plays a part. But until the woman with a woven basket full of dripping faecal matter on her head truly becomes a figure of the past, that compromise cannot be negotiated. 'Total sanitation' cannot be reached and should not be pursued without at the same time insisting on the liberation of the manual scavenger. Whether under activist pressure, public-spirited leadership or as a result of incentives for 'total' toilet coverage, local and municipal leaders should strive to transform the sweeper's work, life and status.

The creation of a new economy around mass toiletization demands that their indignities, as well as those of their customers, be brought to an end.

India's manual scavengers are the best-known, most numerous and most stigmatized of all informal sanitary workers. But they are not the only cadre of their kind continuing the work of traditional night-soil carters until the present day. A rare anthropological essay on the subject, published in 1998, describes how, among the urbanized Akan people of south-eastern Ghana, people known as *Kruni* have for generations undertaken the collection of excreta from households, in exactly the same way as sweepers in the Indian sub-continent. The *Kruni* people originally came from the north, from Sierra Leone and Liberia, and ended up doing work that no Akan would touch. Once a week, invisibly in the night, they visit the households on their roster, take the bucket from its niche, empty the contents into a container, and carry this on their heads to a dumping area at the edge of the town. Their only other equipment is a broom and a lantern to guide their way. *Kruni* are hired by the local authorities and paid around US$30 a month; the household fee for their service is US$0.50.[13] This pattern may also have been common – and still continue – in other long-urbanized parts of West Africa and the Middle East, but the literature is silent. In south-eastern Ghana, the profession is dying out because no-one will take over the work from their fathers.

For reasons which (if the commentator on the *Kruni* is to be believed) are at least partly to do with anthropological researchers' olfactory distaste for surveying traditional facilities, such workers were to all intents and purposes invisible to the international purveyors of public health until very recently – and to many, they still are. Paradoxically, it was only when these lowly operators started to be threatened with loss of work by international insistence on the privatization of water and sanitation utilities that their entrepreneurial activity began to be positively noticed.

Until the 1990s, the assumption prevailed that the delivery of water and sanitation services in any industrializing society should be by public utilities, at public or publicly subsidized cost, on the basis that these were services that met essential needs of life and promoted the public good, and that their provision could not be left to profit-hungry entrepreneurs. Before the municipal authorities stepped in towards the end of the 19th century in Britain, private provision had also proved inefficient at sanitation in the poorer parts of town: local water companies did not want to build sewers and pipelines through run-down and impoverished neighbourhoods for all the obvious reasons. To drive such companies out, state-owned and -run utility operators were set up with monopoly rights to serve all the customers within their jurisdictions – not only in Europe, but in the US, and in client and colonial states worldwide. As a result, bootlegger water providers and sewage removers, even in the sizeable areas of modern developing world cities where nothing else was on offer, tended to be regarded – where they were noticed at all – as racketeers charging usurious rates.[14]

In some instances, water vendors in slums did – and do – charge considerably more per litre than service fees paid by middle-class homes with mains connections; but this is as much a product of utility mismanagement and political failure to charge cost-efficient rates to those who could afford them as of unregulated exploitation. Nonetheless, the nefarious doings of vendors were frequently cited in calls for public utility reform. The inadequacies of the utilities – bad operation and maintenance (O&M) records, failure to set tariffs sensibly or collect them, and their inability to reach residents in lower-income areas – fuelled a major thrust for their drastic reform in the early 1990s.[15]

According to the World Bank and its allies in the international community, the answer to the utilities' failures was their privatization. Partnership with international corporations would bring in much-needed external capital; the market efficiencies riding on its coat-tails would fix the pipes, generate higher revenues and enable the spread of services into poorer neighbourhoods. This championship of the market as the answer to poor public utility performance coincided with the new emphasis on 'demand-driven' approaches – an ideological confluence which ultimately proved unhelpful to the continuing effort to transfer low-cost, on-site sanitation technology into the commercial world. But this is to run ahead of the story.

As the 1990s progressed, water utility privatization began to be imposed upon countries as a condition of structural adjustment and debt-relief packages. Before long, this became an international *cause célèbre* and garnered criticism from many directions. For a while, some much-publicized private sector partnerships (PSPs) managed to extend water connections to poorer urban populations in the Philippines, Argentina and elsewhere. (The services did not include sanitation but at least water connections offered opportunities for better hygiene and cleanliness.) Unfortunately, within a few years, most proved unable to continue to deliver on their pro-poor service delivery promises; companies began to withdraw from municipal deals and even to run fast in the opposite direction.[16] The reason was simple. In order to pay for service extension, it was necessary to hike tariffs to profitable levels in better-off neighbourhoods; but when the companies tried to do this, it proved politically impossible. Municipal authorities baulked, particularly when the service was still very inadequate in the view of the customers and price rises of several multiples were involved. At this point the international partner typically withdrew, citing debts and broken agreements. One way or another, the economics – and the politics – of laying on water and sanitation to the poorer inhabitants of urban spaces, let alone to shanty-towns and squatter settlements, via ventures tied to international currencies and corporations, could not be made to add up.

In the process of this discovery, studies were undertaken that were very revealing about how people living in different kinds of poor urban or semi-urban neighbourhoods in different parts of the world were managing to meet their cleanliness and excretory needs. There turned out to be systems of petty entrepreneurship around water and waste whose service contribution had previously been ignored. A 1998–1999 study undertaken by Tova Maria Solo, an urban planner in the World Bank's Water and Sanitation Division, found that 50 per cent of urban inhabitants in Latin American cities depended on small independent (she refused to call them 'informal') providers for sanitation, and in Africa, the proportion rose to 80 per cent.[17] Solo was one of the first people to try to repair the reputations of these water and sanitation entrepreneurs, previously painted in the blackest of hues. Their ranks included water vendors, standpipe operators, water kiosk caretakers, sludge or wastewater haulers, latrine pit or septic tank emptiers, laundry-people, users of wastewater for fish-pond cultivation and urban farming, rubbish or solid waste collectors, street cleaners, and others who – like India's manual scavengers – have since time immemorial survived from the proceeds of dirt, hygiene, garbage, cleaning and personal grooming services. This too was a 'private sector', but one not taken into account by the ideologues of market capitalism and utility reform. It was consumer-driven, its operators – unimproved and unregulated though they were – managed to make a living, however shitty in terms of occupation and financial reward. And sometimes they were able to hold customers to ransom with their fees because there was no alternative public service.

One of the cardinal virtues of the 'alternative' private sector, according to Tova Solo, lay in its ability to provide services stemming from and adapted to local circumstance. This characteristic was in direct contrast to the operations of the utilities and their corporate friends, whose concessions and contracts were defined by terms of reference established in industrialized world offices. The contrasting living conditions of residents living in towns and cities in the developing world – anything from billionaire suburbs to squatter shanty-towns –

required an anarchic range of potential sanitary and waste disposal response. Monopoly providers with unified solutions could not satisfy the requirements of such varied populations. Her prescription was, roughly: 'let a thousand sewers bloom'. Solo's favourite example of petty entrepreneurship in sanitation was that of a resident of Malang in Java, Indonesia, Pak Agus Gunarti, driver of a *bemo*, a small public bus. Pak Agus' disgust with the poor condition of his neighbourhood prompted him to design a small-bore sewerage system purely with the aid of engineering manuals. He managed to obtain a treatment site for sludge next to the local cemetery, and sold the idea to his neighbours. The installation of the system was self-financed, locally operated and maintained, and served around 70 households. Pak Agus was promptly hired by the City Sanitation Office and invited to replicate his project in 10 more localities.[18]

One of the unforeseen problems with PSPs was that they could not compete with the informal service providers already on the block. And the counterpoint to this was that, instead of incorporating their services and upgrading their skills, the new corporate utilities threatened the informal service providers' livelihoods. This came to light in a study carried out by WaterAid Tanzania in 2003 into the situation of Temeke, a large unplanned shanty-town area in Dar-es-Salaam.[19] This study was part of a wider enquiry by WaterAid into the impacts of proposed or actual PSPs on delivery and spread of services for the poor, undertaken because so much was being claimed on behalf of this route to expansion of services and MDG attainment.

In Temeke, proper garbage collection and engineered sewerage or drainage systems were unknown. Nearly 90 per cent of the 200,000 households (around 1.5 million people) used simple pit latrines; those without latrines used the open air. In the rainy season, when the water table rose, pits overflowed into the lanes and streets, creating a filthy mess and high risks of diarrhoeal disease. To deal with these problems, small-scale entrepreneurs provided a semi-institutionalized solution. These were known as the *vyura*, Swahili for 'frogmen'. These private operators made a business out of emptying facilities located in places inaccessible to pit-emptying trucks. The *vyura* worked in groups of two to four; they began by pouring a solution into the overflowing toilet to kill the stench, and then removed the content by bucket. The sludge was buried in a hole dug nearby for the purpose. They received around 20,000 Tanzanian shillings (US$16) for each pit emptied. This compared with the city's private truck operators, who charged around 25,000 Tanzanian shillings.

The study concluded that it was unlikely that the Temeke population could gain the better functioning services they needed from the PSP. The prospects were not good even for water, since there had been far too little consultation with the inhabitants, and very few resources had been budgeted for the necessary pipelines. And as for *sanitation* in Temeke: this was not even considered in the PSP contract proposals. The conclusions of the wider WaterAid report were that neither the interests of the poor, nor their potential contributions, were being elicited or taken into account in PSP activities generally – very similar problems to those they had faced before PSP rode over their horizon.[20] All the emphasis in PSPs was on the terms of contracts with the private sector – which naturally did not include such informal entrepreneurs as the frogmen. The failure to consult communities about how to solve their sanitary and water problems, or consider their existing solutions as a basis for something better, meant that the technologies proposed were too expensive; lack of state control meant that construction quality and maintenance were weak; and without community ownership, prospects of sustainability were nil. Worse, if the PSP enterprise failed and the private partners withdrew, the communities in question might be left with a costly service and facilities that didn't work; meanwhile the small-time entrepreneurs who had managed things in the past, however imperfectly, would have been pushed out of business and disappeared from the scene.

Gradually, as the limitations of involvement by the corporate sector and PSPs began to emerge, the role of the 'alternative private sector' in sanitation began to gain not only recognition but positive approval. Perceptions swung through 180 degrees, from demonization to active embrace of what were now labelled 'small-scale providers', just at the time when water and sanitation services became regarded as ideologically pure only if they were demand-led, subsidy-free and able to recover their costs. In the past, these small-scale providers had been depicted as only able to offer short-term, and therefore 'unsustainable', solutions. But actually, if they were as pervasive as turned out to be the case, how could what they were offering be 'unsustainable'?

A WSP study into 'independent providers' of water and sanitation in African cities revealed that 70–90 per cent of households dealt with their own sanitary needs. Either they built pit toilets or septic tanks, or they hired people to do so for them. And in the case of the poor, the proportion rose to 100 per cent.[21] The frogmen of Dar-es-Salaam were not unique: in Dakar, the same cadre of manual workers was called *baye pelle* or 'old shovel men'. If people could not afford the *baye pelle* or his equivalent in Nairobi, Addis Ababa or Ouagadougou, they dug out the sludge themselves. Since there is often nowhere to put the contents of their pits, it often gets left in the streets at night – as in pre-19th-century London. Some deliberately let their pits flood in the rains, seeing this as the season of flushing. So the problem is not inability to procure a livelihood out of shit, or unsustainability of the service. The problems are standards (of construction and maintenance) and risks to public health of pit-emptying operations which do not thoroughly and hygienically remove the excreta from the neighbourhood and put it somewhere safe. For example, in Touba, a Senegalese town of 80,000 that swells to over 1 million at the time of the annual Grand Magal pilgrimage, the mass ejection of faecal matter into the streets during the festival typically leads to an outbreak of cholera – no matter how ferocious the authorities' preventive measures.[22] The sudden inflation of Touba's population once a year and its temporary overproduction of excreta represents an extreme situation, but the unregulated nature of pit-emptying where sanitation is uncontrolled is notorious in much of sub-Saharan Africa.

Now that this 'independent' or 'small-scale provider' sector in sanitation has been discovered, and found to be making a living out of muck without recourse to subsidies, donor favours, loans or monopoly rights to customers, the question is: How is it to be encouraged without jeopardizing public health on the one hand, or permitting gross exploitation – of workers and sometimes of customers – on the other? This is a tricky question, and on the basis of experience so far, there are no easy answers. Sanitation services are being provided to a greater extent than previously realized, but mostly by unregistered, non-tax-paying, small-scale masons – *fundis* in Swahili, *mistris* in Bengali – who make their living from unregulated construction, odd jobs or repairs, and count their profit margins in minute denominations. These 'providers' and 'manufacturers' do not exactly look like a sound business investment, quality is not their second name, and many would not be able to read the terms of reference on a contractor agreement nor comply with a regulation if they ever came across one. They know little about public health and do not market themselves on this or any other conceptual basis. Many deny their sanitary occupation if asked, so stigmatized is the trade of shit-shoveller not only in the Indian sub-continent, but everywhere in the world.

How, then, are donors, municipal bureaucrats and engineers to lower their gaze from the spectacle of transnational capital and public utility expertise, and instead start supporting the mini muck entrepreneurs? This is against every commercial and public health principle they were brought up with, requiring a cultural revolution more profound even than the onslaught against the Great Distaste. Nevertheless, if the MDG, never mind the ultimate goal of universal sanitation is ever to be reached, this is one of the new frontiers that will have to be opened up.

If we turn our minds back to some of the many programmes visited during the quest for a new sanitary revolution, it will be recalled that creating a new sanitary workforce – at least for construction – has been implicit in many case stories. In Lesotho, and in Ziguinchor, Senegal, masons were trained to make VIPs on a commercial basis; in Maputo, they were encouraged to set up sanplat boutiques; in many countries the twin-pit pour-flush technology became commercially viable at some level, for example in India in the Sulabh enterprise; and nowadays even in India some manufacturers are marketing fibreglass urine-diversion squatting plates.[23] In Central America, China and Vietnam, urine-diverting and composting toilets have generated a new cadre of technicians, even if its ranks are small. Not since the 1980s, when the World Bank Technical Advisory Group developed prototypes, manuals, standards and training curricula for professional engineering institutions, has there been a similar push to transfer low-cost toilet technology into the commercial world. In many cases, today's enthusiasts envisage the creation of 'barefoot' sanitary technicians with artisanal skills.

Slowly, the training of local masons and replication of low-cost designs has succeeded in transferring toilet technology into the local market – at least in some settings. For example, by the late 1980s shops selling water-seal slabs and concrete rings had begun appearing in bazaars all over semi-urban Bangladesh.[24] This was the effect of decades of training masons in the manufacture of simple pan-and-trap toilets, which began back in the 1960s in what was then East Pakistan.[25] These shops did not cater for the really poor, but for those a few rungs up the socioeconomic scale. The Bangladeshi public health engineering department (DPHE) workshops in which the trained masons were originally employed were not expected to function as businesses. But when the idea was exported across the border to Midnapur, West Bengal, it was reconceptualized for a society with a higher level of income and the same instinct for entrepreneurship. Here, sweepers were not expected to empty pits: another would be dug when the first became full. In Midnapur, a new industry and set of occupations, male and female, were built up around the gooseneck-pan-and-slab and three concrete pit-lining rings, priced at around 250 rupees (US$5). These production centres were set up as commercial operations, selling ready-to-install 'home toilet kits'. Importantly, this was an NGO-operated project, not a government programme – at least, not initially.

In Midnapur, low-cost technology was transferred into the rural economy at a level affordable by almost everyone; this encouraged young entrepreneurs, and provided employment for men and, especially, women. Many women became employed at the centres making pans and gooseneck traps, earning between 20 and 100 rupees (US$2) a day. Others were taken on as 'motivators' or saleswomen, earning 20 rupees for every toilet they persuaded a household to buy. This was a useful extra source of income for women: considering that in one year alone (2002–2003), 900,000 toilets were constructed, the millions of rupees passing through their hands was a boost to income, status and the local economy.[26] Training and some start-up finance for (male) managers of the centres were provided, but they had to make a go of the business themselves. As in all entrepreneurship, some did well, others less well. Other parts of the countryside economy were also boosted: the rickshaw *wallahs*, for example, who bicycled about the West Bengal landscape delivering the ungainly components to their new owners (see Figure 3.4, page 90); pit-diggers and superstructure masons; and the credit unions who provided the loan finance.

When these centres were first set up in West Bengal, they were known as 'rural sanitary marts' and were supposed not only to manufacture and sell toilets, but to act as promotional centres for sanitation and as shops for hygiene and health-related products such as soap, brooms, buckets, domestic water filters, footwear, toothbrushes, bleaching powder and oral rehydration solution for treatment of diarrhoea. In Midnapur, however, the 'drug store' role fell away as the more profitable production of toilets took off and absorbed the attention of managers and workers. Although rural sanitary marts became a central component of the GOI/UNICEF 'total sanitation package' in the early 1990s, outside West Bengal few managed to fulfil such ambitious intentions. Many ended up selling drugstore items and a few ceramic pans, and did little to promote the transformation of sanitary behaviour. Without simultaneous promotional activity on the ground to create and nurture demand, it was not possible for a sanitary mart manager to achieve so much alone. Few attracted low-income customers or extended the retail market.[27] Nonetheless, although many sanitary marts were based on an over-optimistic assessment of existing consumer sanitary demand, they were an important conceptual development. This was one of the first times in a government-backed sanitation programme in low-income areas of the developing world that toilets were promoted via shops as items of household improvement, rather than installed in homes as aids to public health.

The model was exported to West Lombok, Indonesia, to help the 'latrine *Bupati*' with his campaign; here, the women's organization, the PKK, set up and ran the production centres, which were almost identical to those operating in West Bengal. Some sanitary marts are still flourishing in other parts of India, where conditions are right. A few years ago, the model was imported to

Madagascar by Edwin Joseph, an Indian Catholic missionary working with the French Frères Saint Gabriel in Toamasina, the island's second largest city on the east coast. Father Edwin, who was the first person to introduce an improved pit toilet into Madagascar, visited sanitary marts in India and has managed to transfer the whole idea very successfully, on a small scale initially but with every prospect of replication.

The sanitary mart experiment, with all its ups and downs, illustrates how the transfer of gooseneck technology up to a point of commercial viability required resources from a public-spirited source – or international donor – that made it possible to take the risks, iron out initial snags and setbacks, and establish the mechanisms for take-up on a larger scale. It is true that this can also happen without such support, courtesy of the power of the market, intrepid risk-takers and 'mini-venture capitalists' such as Pak Agus Gunarti in Java. But in economically marginal environments, where the market for consumer goods is precarious and unpredictable, and potential customers have no spare cash and live on a day-to-day basis, such risk-takers are conspicuous by their absence. If no public-spirited body steps up to the plate and smoothes the path of the water filter, rainwater cistern, sanplat or pour-flush toilet, ideas with potential for improving people's lives may languish in the wilderness. Putting safety nets under life-improving products in marginal economies is not about promoting 'dependency' or 'supply-driven' approaches, but about trying to forge links between the modern economy of manufacture and consumerism and the traditional economy of near-subsistence.

Due to the Great Distaste, the market economy of the on-site pit toilet has needed a lot of extra assistance, independent of the 'public good' and 'public health' case to be made on its behalf. And this assistance has usually come not from global capital or private investment of the commercial kind, but from another avatar of the 'private sector', the NGO in its vastly different forms. They often receive financial support from enlightened larger donors; but it is NGOs, particularly the smaller community-based organizations (CBOs) whose roots and financial modus operandi belong in the community, that provide conduits from the formal to the informal economic system, and from the world of laws and administrative systems to that of invisible lines of authority and transaction. These NGOs and CBOs occupy the institutional space between the metaphorical world of mud and wattle and that of concrete and mechanical aids. In order to build new cadres of workers and new modes of entrepreneurship around hygienic toilets and excreta, their involvement is almost invariably needed.

An illustration of the NGOs' role can be found in Kibera, an unplanned township on the outskirts of Nairobi. Kibera's 600,000 residents, including many of the city's workers, live not in 'houses', but in what the government terms 'structures' – shacks of mud, timber refuse, flattened tin, cardboard and plastic – because they are occupying the land 'illegally'. This does not stop owners of the structures from making a tidy living in rents, but the fact that they are under constant threat of demolition acts as a disincentive to any form of improvement. Each room covers around ten square metres and provides living space for at least three inhabitants. No environmental sanitation services of any kind – wastewater drainage, public toilets, water standpipes, rubbish collection – are laid on by Nairobi City Council, and none of the lanes are paved. The whole bustling township, full of life and vigour though it may be, is full of refuse and filth, and becomes a mire whenever it rains. The load of sanitation-related disease is considerable: diarrhoea, skin infections, TB, typhoid and malaria.

Over the years various NGOs have attempted to improve the state of sanitation in Kibera, and the growing presence of public and private pit toilets is the result of their prompting. This sanitary effort, inspired by public health concerns, has led to the creation of a workforce in Kibera running 'alternative' sanitary services. These workers include caretakers, builders, pit-emptiers and employees of enterprises running trucks that evacuate pit contents mechanically (Figure 6.2). Some of these earn more than the minimum wage of a manual worker in Nairobi.[28] The variety of entrepreneurial activity in Kibera is instructive, as is the degree to which NGOs and CBOs are lubricating the sanitary economy operated by the 'small-scale providers' (SSPs) by virtue of the employment opportunities they offer. In the case of 20 public toilet blocks, all but one were constructed with NGO or other donor funding. The facilities are managed by CBOs, either on an entirely commercial basis – they hire caretakers and pay them a wage – or by volunteers. Where the use of the block is run as a business, the standard of maintenance and cleanliness is better. In the case of the one multi-use block built entirely at a landlord's expense, costs were lower because much cheaper materials and lower standards of construction were used. Arrangements are similar with toilets built for one household or a small group of households. Construction is partially or fully paid for by the donor NGO, and management entrusted to a CBO. This CBO rents out the toilet to a user group, and provides maintenance services including pit-emptying. In cases where the NGO provides a relatively small share of the financing, usually less overall is invested, the facility costs less (because it is made of flimsier materials) and standards of maintenance are also lower.

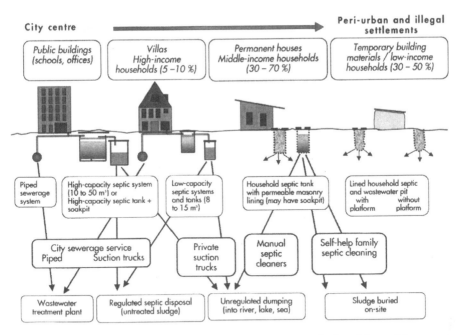

Figure 6.2 *How the urban sanitation market works in African cities*

Source: Bernard Collignon and Marc Vézina (2000) *Independent Water and Sanitation Providers in African Cities: Full Report of a Ten-Country Study*, WSP with other partners, The World Bank, Washington, DC

In one NGO programme, toilet recipients received a grant of 75 per cent of the cost of construction, and in return provided the land and paid the workers' wages. But to qualify for the subsidy, the householder or user group had to use builders trained and certified by the NGO in question. This meant that a group of local artisans became brand-leaders in toilet construction and able to charge higher fees. Thus different donors make different decisions about the balance between external inputs – properly designed blocks using durable materials, public health standards, skills development and capacity-building – and existing community resources and inputs of a less regulated, more informal kind. Thus it would seem that private construction and management of toilets is economically viable in the settlement, but that, unless there is an external input in terms of cash and skills, standards of construction and servicing, and hence of public health, are likely to remain low. Transforming SSPs from bootlegger 'illegals' or cheap casual labour into a respectable and recognized public health engineering workforce remains a challenge.

It is not easy for engineers who have spent their careers spreading water and sanitation services into impoverished slums and villages to start thinking in terms of job-creation, employment, entrepreneurship and toilet-shopping when their whole orientation is towards public service provision for the public good. In Nicaragua, Rafael Díaz, an engineer who until his retirement worked in countries all over the world, has recently undergone this Damascene conversion:

> *All those years, we expected the water and sanitation facilities we installed in poor communities to be managed voluntarily. Now I am convinced that we need people who can make a living out of this, and who are politically and administratively savvy. Only when you have a system in which the community levies fees, sends out the bills, reads the meters, buys spare parts, hires technicians to make repairs and fires them if their work is shoddy, will you have real community ownership and management.*[29]

In 2004, Díaz began his retirement project: the training of 'enterprising community builders', masons who would be able to install water-seal pit toilets, hand-washing and laundry stands, showers, wastewater and stormwater drains, water butts and water filters, and other kinds of domestic improvements. Several courses have since been held in the Matagualpa region, in association with the ENACAL community water and sanitation programme. Trainees, who need already to have some building skills, were identified from villages where ENACAL construction projects and organizational activities are ongoing (see Chapter 3). The course requires that candidates have to pass a test and gain a certificate of competence. Díaz and ENACAL have now run it several times, and the curriculum is in the process of being formally adopted by the National Technological Institute. Díaz has also produced catalogues with drawings and specifications of all the products students learn to make and a theoretical text for them to refer to. Graduates with entrepreneurial flair and sufficient literary and numeracy skills can build on these ingredients.

This is a difficult rural environment in which to build a trade. Many villages are tucked away in remote areas and are usually too small and too poor to sustain independent artisanal workers. One of Díaz's most determined trainees is Jerónimo Valverde, a farmer from a village in San Isidro. His village, Llano del Boquerón, contains only 48 households, and it takes an hour on horseback for him to reach a road where he can pick up a bus into town. In 2004, after he had finished his training, he spoke at one of the regular meetings of the village. 'I told them, I am not a professional mason, but I am here to support my community. If someone wants a toilet or a water cistern, or to line the well, I am ready. But you have to pay for my services.' He gained an NGO contract, and some private commissions, to construct rainwater-harvesting cisterns; to keep the price down, he allowed people to pay for labour with food. He has also made and sold water filters, with the help of his wife. He invested his earnings in his horse and in tools, and is now available for hire as a carpenter too and offers his services in other communities. Valverde has yet to obtain orders for toilets: at present, given water shortages in Llano de Boquerón, cisterns are people's priority and they have yet to appreciate the virtues of sanitation. But Donald Martine, a trainee from a larger community closer to town and with higher disposable incomes, expects to make a good living from toilets. 'When I and my wife have our own water-seal, others will see it and want one too.' Many trainees' main source of sanitary income is from NGO-sponsored toilet-building – NGO programmes are ubiquitous in rural Nicaragua; but at least the fees are now entering the local economy, rather than going to outside builders brought in for the purpose.

This programme assumes, as with the rural sanitary marts in India, that the new entrepreneurs will generate demand for their services. However, in many parts of the developing world, including Nicaragua, this can present problems. People often enrol on training courses run by government or NGOs in the assumption that work and jobs will subsequently flow their way without effort on their own behalf. Alternatively, they see NGO programmes as sources of community investment, and seek out their patronage as the only way of improving amenities and earning money other than by agricultural work. 'Self-help' and entrepreneurship are novel. People's expectations stem from a long history of being under the heel of forces stronger than themselves, and of paternalistic supply-driven approaches, and from the existence of too many programmes whose entire horizon is defined by the idea of supplying money for building amenities in poverty-stricken communities on the basis of household subsidies.

The issue of subsidies and whether they can or should play any role in demand-driven programmes remains a vexed question. In rural Nicaragua, the need for subsidies to lubricate demand for toilets, showers, washstands and drainage will continue until people can access loan finance for home improvements. At the moment, communities shop around for NGO mini-grants towards the installation of facilities such as toilets, and add other improvements – washstands, basins, water filters – when they can afford to. The absorption of sanitation and sanitary workers into the local economy, here as in many other settings, would be significantly boosted by the transformation of local credit possibilities for householders. Villagers who are asked express willingness to pay the whole cost of a toilet instead of just the non-subsidized part if there was a system whereby they could access loans. Lack of means for financing home improvements is a major inhibition to the expansion of the consumer market for sanitation in many impoverished settings.

Most of the discussion so far has been about developing a cadre of artisans able to manufacture on-site toilet parts, and promote and sell them, usually with installation services. And that is almost exclusively the type of entrepreneur or 'small-scale provider' that donors and NGOs have engaged with. The other principal workforce, the workforce that this chapter first engaged with, those who collect excreta from buckets and pits, and who perform work equivalent to that of the sewer, rather than excavation of the pit, construction of the toilet house or fabrication of the toilet pan, has been neglected.

Leaving aside what happens to pit 'sludge' after it has been taken to a common resting-place or been added to the content of the sewers, all users of single-pit toilets in confined urban living spaces need emptying services. In rural areas or in the larger compounds of some semi-urbanized or peripheral town dwellings, there may be space to dig a new pit when the first one is full. But in the majority of congested settlements, the pit can only ever be a temporary storage facility for its contents. 'On-site' sanitation is, in fact, poorly named, because disposal sooner or later will have to be 'off-site' in many cases. The regular removal of excreta or sludge is an absolute condition of the continuing use of the toilet. If emptying needs are not catered for, recent converts to toilet use will lapse back to open defecation or 'wrap and throw'. In studies conducted in southern and eastern Africa, three-quarters of residents reported major difficulties with overflowing pits and lack of emptying services.[30] Yet in virtually every sanitation programme and in most of the literature about them, the subject of how to ensure operation and maintenance for on-site toilets has been neglected.

In India, those active on behalf of manual scavengers are, for the most part, unwilling to contemplate organization of the workforce or upgrading of their work as earlier explained. In other settings where frogmen or *baye pelle* do not merely remove solid waste or clean the streets but handle raw faeces by emptying pits, they too suffer from the foul, unhealthy and stigmatizing nature of their work, although since they are not bound to it by caste or birth, they suffer less. One of the earliest efforts to provide decent pit-emptying services in areas unreachable by trucks was undertaken by a Dutch NGO called – appropriately – WASTE. At the invitation of the then Dar-es-Salaam Sewerage and Sanitation Department (DSSD), WASTE set out to introduce a semi-mechanized manual pit-emptying technology (MAPET) into the labyrinthine lanes of low-income neighbourhoods. They wanted to develop an economically viable sludge extraction service in confined spaces, using the kinds of pumps and engines with which local workshops were already familiar. Getting frogmen out of the pits and promoting their chances of self-employment by upgrading their skills and entrepreneurial prospects, and linking these into public sector service operation (the DSSD was then still a centrally owned and operated utility) was the principal WASTE target.[31] Their handcart service, with a tank size of 200 litres, pumped out pits load by load and dumped the sludge in an adjoining pit dug for the purpose. By the end of the pilot phase in 1992, seven teams were active in Dar-es-Salaam. The service was successful in many ways: it was cheap, popular and much more hygienic than traditional emptying; it was also in keeping with the informal, small-change economy in which slum people lived.

Over the course of the next few years, however, these services gradually collapsed. One problem was the dependence of the MAPET pump-and-dump machines on spare parts only available from abroad, and the difficulties of repairs and capital replacement. Other reasons were not within the operators' control. The DSSD was dissolved during the privatization of water and sanitation services in Tanzania, and there was no further institutional support; responsibilities for environmental health services were fragmented, and the whole issue of sanitation in low-income areas was downgraded and treated as an adjunct to water supply. In addition, settlement had become much denser as the city's population swelled, and many areas were low-lying and subject to frequent flooding. Both these factors made digging extra pits for on-site disposal impossible.[32] WASTE had started to provide a system of carting sludge to depots where it could be picked up by trucks, but this languished with the extinction of the DSSD. This experience made it clear that a pit-emptying service for crowded and flood-prone areas was urgently needed, but that this has to be accompanied by other activities, which will almost certainly have to be subsidized and underpinned by the public sector. While the household can be expected to install the toilet, and meet some or all of the pit-emptying costs, the water and sewerage authorities will have to provide the means for transport to the sewage plant for treatment and for final disposal of sludge.[33]

An evolution of the MAPET technology was the 'Vacutug'. This was another handcart-sized machine, originally developed in 1996 by an Irish engineer, Manus Coffey, for use in Kibera at the request of UN-Habitat in Nairobi (Figure 6.3). The Vacutug had a 500-litre capacity and was capable of emptying eight toilet pits a day; the price for its services was around US$8.[34] Greeted with the enthusiasm that such intermediate technology devices often attract, and piloted by the

Figure 6.3 *The Vacutug developed for use in Kibera, Kenya*

The Vacutug is a scaled-down version of a vacuum tank pit exhauster, made for operating in small lanes and camps where no large vehicle can pass. Its tank has a nominal volume of 500 litres and the pump/tug assembly has a 5.9kW petrol engine. It moves at 5 kilometres per hour, and when at work fills the tank in around two minutes.

Source: The Vacutug Development Project, UN-Habitat, Nairobi

Kenyan NGO KWAHO (Kenyan Water for Health Organization), the Vacutug has been under trial ever since. Only around 10 have ever been built, for use in Nairobi and in a few other countries. With all the technical difficulties of deployment in rutted and muddy lanes, and the lack of a collection and treatment infrastructure, not to mention the cash-flow problems of Kibera residents (which means they tend only to have removed the more fluid content in the top metre of the pit), the Vacutug service was bound to face teething problems. But the failure to progress further in more than a decade underlines the lack of urgency associated with sanitation services in low-income areas, and the eclipse they suffered during the period when international donors were fixated on utility privatization and private sector partnerships as the key to service spread. If this was a new water filter, handpump, small-bore drilling rig or rainwater-harvesting device, within the space of 12 years there would have been more R&D money, vigorous and rigorous trials, company tie-ins, alternative models, capacities and sizes, manuals and standards, and marketing promotion.

In Kibera, after all these years and in the face of bureaucratic inertia, a handful of operators are now managing to make reasonable income from the

Vacutugs they are running, but so far without managing to recover the capital costs of the equipment.[35] Despite this discouraging record, however, technologically simple, cart-borne emptying services, and adjustments to pit toilet design to facilitate the process, remain among the most promising and important sanitation developments for low-income urban areas.[36] In 2000, WaterAid Bangladesh imported a Vacutug and adapted it for use in the narrow lanes of Dhaka. Operated under the auspices of an NGO partner, Dushtha Shasthya Kendra (DSK), this enterprise seems to have been better planned and administered. Over time, the demand for its use in slum neighbourhoods has grown. The operators charge around US$3.50 to remove 500 litres of sludge, and they have the municipal authorities' permission to discharge their loads into the sewers. Local sweepers, who earn part of their livelihood from pit-emptying, are given a 10–20 per cent commission on every order for Vacutug services they bring in. However, there is still some distance to go before this service becomes a viable entrepreneurial concern – capital costs so far having been borne by WaterAid.

According to a 2004 evaluation of the DSK operation, the enterprise needs two important assets: more forceful publicity and marketing, to make people aware of its existence and services, and solid political commitment.[37] Without the latter, it will not be able to develop the necessary linkages and partnerships with local government corporations and agencies for institutionalization and expansion. This seems to be the most difficult barrier to surmount in terms of engaging the public health authorities with the needs of slum areas: they are happy to allow NGOs to promote the construction of toilets in the slums, but they are not willing to establish a service infrastructure to support them effectively nor to undertake the necessary excreta transport and treatment services themselves. The question of how to reach the poorest households and neighbourhoods also persists. At present, DSK operates the service in such a way as to cross-subsidize its use by the poor; if the service was entirely taken over by private entrepreneurs who were not subjected to effective controls and regulation, the poor – for whose benefit the Vacutug was invented – would lose out. Here is another example of how difficult it is to introduce and economically maintain a semi-mechanized service for consumers whose livelihood margins are too narrow to support its establishment.

Meanwhile, another simpler and cheaper device for evacuating pit contents – named a 'gulper' by its inventor, Steven Sugden – is undergoing trial. The gulper uses a direct-action handpump and wheeled or portable containers to receive and transport the sludge.[38] This approach is an effort to upgrade the shovel by improving the lifting process, rather than an attempt to miniaturize the tanker with its mechanized fuel-driven suction pump. The physical separation of the lifting device from the sludge container makes the operation more flexible: closed bins of shit can be towed away or put in the back of a pick-up, while the gulper continues to operate at the house of the next customer. Over the course of the next few years, it must be possible to develop and transfer basic pit-servicing technologies into the micro-enterprise arena, and – with loans and licensing arrangements – kick-start entrepreneurial activity around something better than shovelling shit by hand. Donors and NGOs are getting involved, but much too slowly. The Dutch NGO WASTE is one of the few trying to unlock the financing puzzle and reduce other obstacles to the growth of a market economy around rubbish, shit and recycling of wastes in low-income parts of the developing world.

The close involvement of NGOs and CBOs in bringing water, sanitation and waste disposal services into communities reflects the importance now attached to people's participation in selecting, planning and siting new facilities in their pre-construction phase, and operating and maintaining them afterwards. For the O&M functions, without which toilets become unusable and fall into disrepair, and water standpipes or pumps break down or create standing waste-water problems, a range of management and technical skills are required. Those committed to community-based solutions to power the new sanitary revolution expect these tasks to be handled by local water and sanitation committees, along with caretakers and cleaners who are sufficiently motivated to do their work for a tiny stipend instead of a living wage. They look to the development of local savings clubs for spare parts and replacement facilities, and to small-time technicians to carry out minor repairs.

By contrast, those who see the development of a home improvements consumer market as the principal way forward tend to think more about how to build demand so that appropriate technology of all kinds – toilets, pit-linings, emptying systems, composting, biogas digestors, sludge treatment plants – is transferred into the non-sewered environment in such a way as to generate jobs, products, incomes and satisfied customers. They are not opposed to community initiatives, but on the whole they see sanitary installations as a matter of individual household choice, and they are trying to promote demand and reduce the gaps between wanting a toilet and being able to satisfy the consumer instinct. In this ideological scenario, subsidies act as a disincentive, artificially restricting the prospects of a demand-led sanitary boom among low-income subscribers.

There is no doubt that bringing an on-site sanitary economy into being on a significant scale is a difficult challenge from whichever direction it is addressed. But if the whole of the sewered and industrialized world has had their excreta removal system – not their toilets, but the removal and treatment of their wastes – subsidized from the public purse, why is it sensible or fair to demand of the poorest people on the planet that they pay for the whole operation themselves?

Before the era of privatization, it is difficult to picture any enthusiast for sanitation arguing that services should fully recover their costs – capital and O&M – from users, especially when it comes to users who are poor. The experience of the 19th-century sanitary revolution showed that when provision of services was left to the private sector, they were confined to those neighbourhoods and customers that could afford them. There was no reason to believe that the experience would be different in the confused and irregular economic environment of the developing world. But, as noted earlier, the championship of the market as the way to eliminate the inefficiency of state-owned utilities coincided with that other turn-around: the conviction that demand, not supply, should drive the provision of sanitation and water services.

Only when people, even poor people, showed by their willingness to pay that they truly wanted new facilities and services could adoption of sanitation, and the hygiene behaviours surrounding it, be sustained. The outcome of this ideological confluence was unfortunate. Instead of working out how to reform public utilities so that their performance was stronger and their ability to reach customers in lower-income areas improved, the simplistic formula of market efficiency and privatization neglected sanitation for poor people and left them looking after their own needs. Many enthusiasts for the 'demand-led' approach unwittingly cooperated with this abandonment of the rights of poorer people to sanitary provision, on the basis that if you did otherwise you created dependency, distorted the market, deprived incipient small-scale providers of their livelihoods, inhibited community awareness, and crushed people's sense of ownership and self-respect.

It is reassuring from the point of view of gauging demand to learn that private purchase was far more important than government and donor investment – at least US$26 billion in private cash, as compared to US$3.1 billion in public money – in enabling around 1 billion people in Asia, Africa and Latin America to install a basic on-site toilet between 1990 and 2000.[39] Another way to look at this record is to point out that if there had not been such an abandonment of the poor, and public sector investment in on-site sanitation and its accompanying institutional infrastructure had been at an appropriate level, the combination of high demand and efficient supply might have multiplied consumer take-up many times. A salutary lesson can be learned from an earlier story, that of the abandonment of subsidized public housing as part of urban policy in the developing world. This was done in response to the claim by proto-anarchists of the Ivan Illich school that slum-dwellers' own solutions to their housing needs were much more efficient, creative and empathetic than anything public money could provide.[40] These ideas on urban housing were famously expressed in a book by architect John Turner, *Housing by People*, published in 1976. By arguing persuasively – and accurately – that slum people possessed ability, courage and capacity for self-help, those committed to participatory urban self-improvement – and more generally for people's empowerment rather than public investment as the way out of poverty – unwittingly prepared the way for a withdrawal of state and local government support.[41] The costs that slum inhabitants faced – high unit prices for construction materials, substitution of poor-quality items, lack of protection against shoddy work, requirements for community services – were given insufficient consideration, as compared to their creative capacities and right not to have their settlements razed to the ground and replaced by tenement-style housing.

The extolling of slum and shanty-town dwellers' abilities to solve their sanitation and toilet needs – something that has been effectively imposed upon them by lack of government and donor engagement – can be similarly interpreted as a pretext for leaving them to wallow in their mire. In order to reinforce the case that sanitation is best done subsidy-free, NGOs and donors have a vested interest in being able to demonstrate that installing a toilet can cost virtually nothing. Some of the prices cited today for low-cost sanitary toilets – around US$1.25 in Bangladesh, for example – are unrealistic.[42] One bag of cement costs much more than this. Similarly, when poor people are put under pressure not to excrete in the open, but given no financial support for sanitary construction, the short cuts made by local *mistris* (builders) with pit-lining and other materials are celebrated as if some extraordinary tensile powers had just been discovered in stubble and dung. The romanticization of low-cost construction is unhelpful to the advancement of public health, without which the sanitary revolution serves little purpose.

These alternatives to open defecation are not likely to be congenial, odour-free or offer protection against disease in any lasting way. The excuse made for supporting this approach is that once people have put their foot on the first rung of the 'sanitation ladder', they will carry on upwards to the improved toilet when they can afford to – an outcome far from guaranteed. Unless used with care, this proposition can be a useful device for letting the public health engineering services off the hook. Allowing public financing to be cut to the bone and pretending that this will improve service spread is not the way forward. No on-site toilet, if it is to be efficient, health-promoting and nice to use, can be virtually cost-free. There is a real problem concerning the costs of this particular home improvement in environments where people are living in circumstances of serious financial stress, and suggesting that there isn't does not help promote the sanitary revolution or reach the MDGs.

Some of the arguments against subsidies for household toilets do have merit, especially as many of their protagonists are not arguing for privatization and cost recovery across the board. On the contrary, they are arguing for public investments in public health institutions, governance and regulation, marketing, technology R&D, and other underpinning components of a new sanitary order, while ending or reducing subsides for the pan, slab and pit.[43] It is also the case that the strategy of paying households to receive sanitary installations has a very poor record: in India, as described in Chapter 4, the provision of a free toilet without accompanying information and mobilization virtually guaranteed its lack of use. At least as a toilet. And there have been many cases where subsidies have been captured by better-off residents, or better-off communities, not only in India but everywhere. No-one designing a programme today would propose that subsidization should be the predominant strategy for sanitation spread. And equally, except in cases of abject indigence – AIDS victims, elderly widows, child-headed households, those with disabilities and unable to earn – no-one would suggest that a toilet bowl and its various fittings should be provided to any house-hold completely free of charge. But it sometimes seems that, in trying not to replicate the mistakes of the past, 'no subsidies' has been put forward as the latest quick-fix solution, instead of recognizing that subsidies need careful handling and management, with community input and transparency.

Adopting a position of anti-subsidy absolutism is as potentially misguided as any other 'one-size-fits-all' proposition, and ultimately it will not do much to help lift a new economy around the sanitary pit toilet off the ground. There may be urban and rural settings of modest means in the developing world where the doctrine of toilet use has now spread so well or fallen on such fertile ground that freedom from subsidies for construction is viable for the majority of house-holds. There are many more settings where it has not. Among the latter, it may be the case that those who are motivated but do not have the cash in hand can access micro-credit – in Bangladesh, India or South Africa, for example. In other settings, this is not the case: savings and micro-credit institutions are new, and it usually takes years for them to develop. The technology of sanplat or pour-flush may also be newly introduced and production costs still relatively high; until the technology has been properly transferred, there will be insufficient small-scale providers for prices to drop substantially. In many environments, poverty is acute and other day-to-day priorities – food, shelter, water – are so much more important that it is not realistic to expect full cost recovery for sanitation from people at the margins, either for construction or for ongoing O&M (including pit-emptying).[44]

Purists about the evil nature of subsidies would have a hard time explaining to many enthusiastic customers for sanitation – in Senegal, in Kenya, in Andhra Pradesh, in Nicaragua – why a household toilet grant would be such a crushing disincentive. There are in fact very few contemporary sanitation programmes for low-income areas where 'no subsidies' actually prevails. Experience in southern Africa suggests that attempts to recover full costs from people in low-income urban settings simply end up by further marginalizing the poor.[45] In the end, the case for full cost recovery seems to be a theoretical argument for spreading existing resources more thinly in order to reach the MDG, when actually what is needed to improve the quality of sanitary life for millions of people is more resources, wisely applied.

Take the case of Madagascar. When Edwin Joseph of the Frères Saint Gabriel arrived from India to work in Toamasina in 1999, there was no low-cost household sanitation programme in the country. No-one in ordinary rural or urban society had ever seen a sanplat or a pour-flush water-seal latrine. A major inhibition was the existence of strong taboos – *fady* – against the storage of excreta underground (see Chapter 3), which observers believed would stop any effort to promote household pit toilets in its tracks. After three cyclones hit coastal Toamasina within three weeks in 2001, and an epidemic of cholera began to fell whole families in his parish, Father Edwin took up the challenge of breaking the *fady* by instigating a sanitation programme. He started in a community of fishing people living close to the heavily polluted Pangalanes Canal. He involved local cultural and family leaders, and had animators visit every family and fully discuss the idea. The pilot project was extremely successful: far from rejecting the idea of a pit toilet, demand was overwhelming. The project has since spread to 16 suburbs out of 138 in Toamasina, and people from all the other suburbs are requesting its spread to their own. Over 5000 households have installed toilets, and there are a further 6000 names on the waiting list.

So what is the financial formula used by the Frères Saint Gabriel? The cost of a pour-flush toilet with concrete pit-lining rings from the sanitary mart is still high, given its novelty: around US$100. This is a huge sum of money for an impoverished Madagascan family. Many families in Toamasina who want a toilet can afford this, and around 800 have used the sanitary mart on a purely commercial basis in the two years since it was established, but these are not the poorer and disadvantaged customers that the Frères Saint Gabriel are interested in helping. So they conducted a survey in the communities to establish people's ability to pay. A set of criteria for subsidization were then agreed. In the poorest group are those whose housing is of natural materials, who have many children and little income – the household head pulls a rickshaw, for example – or where the householder is a widow or elderly; these candidates pay around US$8. In the second group are those with a tin roof on the house, modest but regular employment and some household amenities such as electricity; these pay one-quarter of the cost. All other customers pay the full amount. Father Edwin believes that the whole issue of *fady* has been much exaggerated. He has had many requests for facilities from rural towns, but he believes that the main way to service these is for the authorities to set up sanitary marts and micro-credit access. In the opinion of the local WaterAid staff, until there is more local experience with micro-credit, it is impracticable to talk of 'no subsidies' for household sanitation for poorer Madagascans.[46]

When it comes to the management of sanitation – pit-emptying, excreta transport, treatment of effluent and final sludge disposal – the argument against subsidy evaporates. The installation of a toilet is principally of comfort and value to the private householder. But the management of excreta – including the smells that rise from the neighbour's next-door plot, the presence of disease-laden dirt on paths and common land, and the removal and ultimate disposal of stinking pit 'sludge' – affects the whole community. The public health and development benefits of good sanitation extend beyond the private benefits gained by the individual who chooses a sanitary toilet over open defecation or 'wrap and throw'; this has been the economic case for subsidy provision for sewerage in the industrialized world, and should apply to on-site sanitation in the developing world, bearing in mind that the on-site facility doubles as a toilet and a sewer, fulfilling both the private waste emission and the public waste disposal requirements. The question of what subsidies should be used for and how they should be allocated, and of financing generally, has to be linked to the socioeconomic and political context. Resources are desperately needed for extending sanitation systems. But as in all areas of developmental change, that does not mean that resources are easy to apply, or that there is any one formula for their use.

Until production of pit-toilet components and installation services have been transferred effectively into the local artisanal and consumer mainstream, dramatic

7

Bringing on the
New Sanitary Revolution

Previous page: In ecological sanitation systems, twin
compartments used alternately allow faecal matter to be
composted for at least a year. Here, a local sanitation officer
in rural Mozambique shows that when the compost is finally
removed, it is completely inoffensive.

Source: Jon Spaull, WaterAid

L ooking back on the 19th-century public health revolution, a number of landmark moments come into view. First up is Edwin Chadwick's 1842 *Report on the Sanitary Condition of the Labouring Population of Great Britain*; then the Public Health Act of 1848; following this, John Snow's 1854 insistence on the closure of the Broad Street pump to prevent the spread of cholera; and finally the 1858 'Great Stink' off the Thames that inspired London's retching MPs to legislate reform. Whichever is taken as the critical starting-point of the public health revolution, it is salutary to note that it took between 50 and 65 years of legal, administrative, financial, technological and promotional combat with 'excrementitious effluvia' to sanitize the people of Britain and make a real impact on their life expectancy and standard of health. If we date the contemporary international sanitary revolution from the start of the Water and Sanitation Decade in 1981, we have barely run half the course. It is also worth noting that we have achieved pitiful results at a similar snail's pace.

Since 2000, however, the pace has been gradually hotting up. The charge was led by those lobbying hard for the addition of a target on sanitation to that on water, under the umbrella of the Millennium Development Goal (No 7) on environmental sustainability. This was achieved at the 2002 Johannesburg Earth Summit. The target was not articulated separately in its own form of words, but added onto the water target in a similar formulation: to halve by 2015 the proportion of people in the world who in 1990, at the end of the Water and Sanitation Decade, did not have 'access to basic sanitation'. (Curiously, the word 'basic' was substituted for 'improved', the word used by WHO and UNICEF to distinguish facilities that are safe and hygienic from those that, generally, are not.[1]) Since the numbers of people without sanitation are far higher than those without drinking water, the target for sanitation in absolute numbers is far lower than that for water. Even if the target were attained by 2015, it would still leave around 1.8 billion people (factoring in population growth) on the wrong side of the lavatorial line[2] – mostly among those who are poorest and most difficult to reach. But, despite this caveat, reaching the target would still be a major boost.

'Access to basic sanitation' could be translated from bureaucratese in a number of ways, and will inevitably be understood differently from country to country. But if the translation is to be meaningful, and include the idea of 'sustainable' as specified for water, it means continuing and affordable use of a toilet that does not stink, collapse or overflow, and that confines raw excreta on a permanent basis, or is managed in such a way that its contents can be taken away and disposed of safely. Meeting the MDG target in a meaningful way requires that, between 2005 and 2015, toilet facilities meeting these exacting standards will have to be delivered to around 1.6 billion new users worldwide.[3] This is at a time when the number of people in sub-Saharan Africa without sanitation (whether 'basic' or 'improved') has actually been rising.[4] It is also at a time of extraordinarily rapid urbanization, with more people daily setting up cramped and ramshackle dwellings in shanty-towns, on pavements and on illegally occupied wasteland. In many such settings, facilities of any reasonable standard, however the word 'basic' is to be interpreted, will not be accessible for some time to come. Much depends on whether the authorities are willing to address issues external to public health, such as settled land tenure and legitimized occupation, without which there is no incentive for occupants or owners to make improvements to housing, including installing a pit toilet for one or a group of houses.

So at the time of writing, there are just seven years left in which to precipitate many policy U-turns, overcome the Great Distaste, not to mention the Great Inertia, galvanize resources at local and international levels, transfer excreta-removal technology, build a new political economy around low-cost systems, transform attitudes and behaviours among potential 'adopters', and accomplish things that, even in the speeded-up world of the 21st century, cannot realistically be done in the time. That is, of course, unless the guardians of the sanitary MDG want a crash programme of pit-digging and toilet-house building, with all the familiar risks of cluttering the landscape with redundant brick and tin-sheet cabins, and pits overflowing with faecal matter with no emptying services at hand. A building extravaganza might be achieved if vast resources were dedicated to old-fashioned supply-led approaches. But it could not accomplish the institutional reforms or the transformation of sanitary behaviour needed to clean up the environment and bring about the better health that a new sanitary revolution – and the MDG itself – aims to achieve.

During the quest undertaken by this book, every programme visited and toilet missionary encountered has discredited this kind of imposed sanitization strategy and shown that it is doomed to fail. This is not to suggest that strong leadership and publicly expressed zeal – of the kind emanated by Chadwick and Snow once upon a time, Gandhi and Pathak in the more recent past – do not have a place. But the sanitation revolution needed today cannot be orchestrated from on high as if it were a military campaign. Sanitation has to capture the imagination of consumers as a life-improving benefit; it is, to borrow the military jargon, a 'hearts and minds' issue, and the machinery to bring it on will necessarily be institutionally complex. This requires a cultural revolution, not only among potential consumers, but among sanitary engineers, bureaucrats and politicians.

The MDG target has a role to play in revving up that machinery. The whole MDG project has refocused global attention on the elimination of serious poverty, helped rejuvenate the UN's economic and social development system, and refreshed the climate in which its organizations and those in non-governmental circles operate. Working to attain common goals helps confer a common sense of purpose, and targets provide understandable benchmarks against which progress can be measured and reported. Those labouring in the international vineyard of filth believe that the MDG sanitation target has had an important effect in raising the international political stakes and generating momentum for sanitary advance.[5] They cite evidence of institutional and policy change at the country level; and at the global level, current shortfalls in progress towards meeting the goal have led to the UN declaration of 2008 as the 'International Year of Sanitation', the first time that sanitation has been decoupled from water and given recognition on its own. The MDG has also opened doors to new financial resources. Therefore every organization involved in public health from WHO to WaterAid, from UNICEF to the Senegal Ministry of Sanitation, from the World Bank to the Frères Saint Gabriel in Madagascar, embraces the MDG target for its potential in pushing the crisis further up local and international agendas and provoking the necessary response.

The target's very existence has obliged policymaking bodies in developing countries and international donors to review their water and sanitation activities. In Westminster, for example, in today's stink-free House of Commons, the Committee of MPs charged with reviewing the work of the UK's Department for International Development (DFID) recently took days of oral and written evidence from a large cast of characters and issued a report criticizing the international sanitation record:

> *Sanitation gets far less attention than water in DFID's policies and this imbalance needs urgent correction. On current trends, the MDG will not be met until 2076. This is a hidden international scandal that is killing millions of children every year.*[6]

DFID is far from unique among official donors – in fact, its record is better than that of many others, as it pointed out in its response.[7] In their report, the House of Commons Committee reversed the usual ordering of the phrase 'water and sanitation' to 'sanitation and water'. Such a change in nomenclature is more than cosmetic: it indicates that mindsets within the donor community may be on the threshold of a much-needed transition.

So far, so good. MDGs can make a difference in all sorts of useful ways. But there is a problem, nonetheless. Since their elaboration, the Goals have taken on a life of their own and accumulated their own international baggage. It sometimes seems as if reporting the progress of this important international endeavour has taken over the development discourse. Simplification of complex issues to measurable goals may be necessary in order to capture public and donor attention, but there is also a danger that it can backfire. Great progress may be made in the next seven years in creating the circumstances for sanitary revolution 'lift-off', while still falling well short of the target. Will disappointment then intervene and donors take flight, while the international sanitation and water community tries to defend itself against the taunt of failure, as it did in the years following the Decade? At the moment, attainment of the Goals is too often projected as synonymous with eradication of the problems they address as if coverage figures are the be-all and end-all in the fight against poverty and ill-health. Decades of experience teach us that, when it comes to improvements in the quality of life of the most disadvantaged people in the world, this can be an illusion.

Indicative as coverage figures may be, the changes that need to take place in order to advance significantly the sanitary frontier are much more profound. The attainment of the sanitation MDG should not be sought at the price of abandoning the vital lessons learned over the past 25 years. Can the target's existence and the publicity and activity surrounding it really facilitate – as some experts believe – an acceleration in the take-up of sensible approaches? Or will the critical corrections needed to so many failed programmes – higher quality, more popular involvement, respect for consumer demand – be just the features abandoned in the effort to push up the numbers? This kind of criticism is already emerging in India, as we saw in Chapter 5, in the context of the 'Total Sanitation Campaign',[8] and it will certainly emerge from other locations in which environmental activists are fewer on the ground and programmes less frequently put under a critical spotlight. Looking back at the earlier Water and Sanitation Decade, it is true that the effort it generated did provide a useful fillip for the process of sanitary discovery and policy review, and many 'new approaches' did come out of it, even if its notional goal – 'Water and sanitation for all' – was over-ambitious. Let us hope that the MDG can and will do the same.

The MDGs for water and sanitation are cast rather differently from the Decade target. They are not stated as noble aims, but as specific time-bound measurable and reportable objectives to benchmark progress towards the ultimate end of 'universal' access. There is a lot to be said for introducing definitions, indicators and other survey tools which make data collection possible – both to quantify how programmes are going and to get a handle on how global numbers are stacking up. But we are beginning to discover that counting the number of households with toilets designated as 'improved' often reveals little about whether they are used, who uses them (male and female, young and old), and whether householders always use them and make their children do so too, or about what happens when people are away from home or when the toilet is 'out of order'. Questions of whether there is full 'access to sanitation' by all members of the household, including those who are sick or disabled or vulnerable in other ways, and about regular and dedicated use of a toilet facility – the questions whose answers indicate whether behaviour has really altered – turn out to be complex. In a study undertaken in Nepal, for example, it was found that two-thirds of the population continued to defecate in the open, although half the population has access to some kind of toilet.[9]

If the local or household convenience serves a large number of people and is frequently 'occupied', or if it has not been recently cleaned or emptied, what do you imagine that many of them do? Even when householders have been convinced and have installed toilets, they may not use them all the time – particularly when they are away from home and working in the fields or elsewhere with no facilities, at markets, for example. A recent study from Cambodia, Indonesia and Vietnam which examined patterns of toilet behaviour within families and communities according to many variables – age, gender, socioeconomic status, occupation, time of day – concluded that 'coverage' is a poor indicator of success in sanitation programmes. The author, Nilanjana Mukherjee, went so far as to suggest that counting toilets was a futile exercise if what the programme was trying to do was improve the standard of community health. She commented that, without the development of suitable products and service delivery options, especially for the poor, 'coverage rates could be quite misleading, while the goals of access and community health impact may forever remain out of reach.'[10] In all three countries, ownership of a household toilet did not imply a consistent change in the household's sanitation behaviour (Figure 7.1). This finding would surprise few of today's up-to-date practitioners and experts; in fact, some make the point that, as yet, we know far too little about toilet usage patterns and how these interact with health to make assumptions about the impact of 100 per cent toilet coverage on wellbeing, including child wellbeing. Expanded coverage is very important, but it is not in itself enough.[11]

A sneaking fear remains that efforts in the next few years could be diverted into pushing coverage rates up in order to achieve measurable advance along the

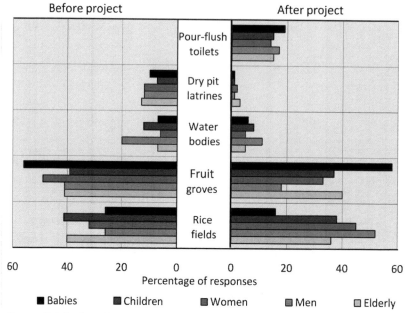

Figure 7.1 Defecation sites used at various times by Cambodian villagers

Surveys undertaken with people in Cambodian villages showed that, even after the installation of improved pour-flush toilets to replace unimproved latrines, usage remained patchy. This especially applied when people were away in the fields working. However, there was a definite shift in favour of the pour-flush, and some reduction in occasional use of fruit groves and water bodies.

Source: Nilanjana Mukherjee (2001) *Achieving Sustained Sanitation for the Poor: Policy and Strategy Lessons from Participatory Assessments in Cambodia, Indonesia and Vietnam,* Water and Sanitation Program – East Asia and the Pacific, Jakarta

MDG-prescribed path. Sustainability, in the sense of abiding and permanent change in service effectiveness, consumer behaviour and public health impact, could be lost if the main focus is on installation. When large amounts of new money become available, there is always a strong temptation for governmental departments and less enlightened practitioners to spend it on construction. It is much more time-consuming and, if not necessarily hugely expensive, difficult in other ways to create and operate a good regulatory and administrative frame-work; carry out local assessments of consumer demand and hygiene understanding; introduce credit mechanisms so that people can pay for home improvements; improve the technology of low-cost pit-toilet construction and emptying services; boost and regularize the operations of small-scale entrepre-neurs, and generally create the market circumstances in which a mini muck economy could flourish. There is as yet insufficient 'good practice' experience in many of these areas to be able to risk short cuts.

Pressure to attain the Goal could undercut this painstaking but vitally important range of activity. More practitioners should be trained who think about the task in hand in a more rounded way, taking the time to prepare the ground and get things right – as did the Ramakrishna Mission in West Bengal and other recent sanitary pioneers. The last thing that needs to happen is that we lose sight of the prize of better health and living standards for the poorer members of humanity – which many project examples and sanitary initiatives described in these pages show as potentially within our grasp. Some of the key experiences examined, both from the earlier sanitary revolution and from activity in the poorer parts of the developing world over the last three decades, have helped to indicate ways forward. Stale assumptions and diagnostic mistakes need to be discarded and replaced by new ideas that, due to taboo or lack of public airing, have not been given enough attention. More information needs to be gathered in many settings before not only the 'why' of sanitation spread, but also the 'how to do it here' can be laid with conviction before policymakers, programmers and consumers.

New approaches should never be advocated as formulaic prescriptions. In the different regions and countries of the world, and within different districts and localities right down to community and neighbourhood levels, the ingredients for effective and sustainable sanitation – technological, institutional, financial, logistical, environmental, social, economic, cultural and promotional – will have to be selected according to different menus and recipes, and put together on the spot. In contrast to the centralized sewerage solution of the conventional kind, a kind of toilet anarchy – horses for courses, pedestals and sanplats for performers – needs to prevail.

Let us start a brief recapitulation of key lessons learned with the Great Stink of London, the phenomenon that, exactly 150 years ago, finally induced the British Parliament to invite Sir Joseph Bazalgette to construct large underground sewers that would collect the effluvia of all the city's inhabitants and transport it to new outlets downstream. The word that needs to be emphasized here is the word 'stink'. For reasons of delicacy, this word – or synonyms such as stench, smell or malodour – do not often appear in today's accounts of sanitation programmes or in advocacy about them. Yet this factor – the smell of raw excreta – is at the heart of all efforts to improve sanitation, as the inventors of the water-seal U-bend and the VIP fully appreciated. Until it is treated, composted or fully dried out, human excreta is a truly noxious substance, and its worst aspect is the smell. If this was not so bad, faeces would still be disgusting, but not quite as difficult to deal with.

The Victorians accepted this – they were forced to. In 1858 the fumes off the Thames became so intolerable that the Great Stink was known by its proper name. At the time, as we noted in Chapter 1, the accepted theory of infectious disease was that it was carried by 'miasma' – in the air. So the Great Stink, caused by hot weather and the reduction of the river to a trickle consisting almost entirely of sewage, was thought not just to be offensive, but to be spreading cholera through the nostrils – an aspect that did much to concentrate the legislative faculties of MPs. Ever since it became understood scientifically that water rather than air is the standard vehicle for the distribution of diarrhoeal pathogens, and gases emanating from open drains known not to be pestilential, the case against 'stink' has taken a back seat. Everyone knows about the unpleasantness of the smell, so why indulge in unnecessary vulgarity? Water and sanitation – though with too much emphasis on water – have gained steadily in reputation as keys to improved public health and rising life expectancy. The reduction of stink or 'bad airs' as a primary object of sanitary progress is no longer so often in the frame, as it would be if the miasma theory of disease had not been discarded. In a way, this is a loss. It indirectly explains many sanitary programme failures.

As we have seen, throughout most of history, people – unless they lived in congested cities, where even in antiquity authorities did something to remove excreta from homes and streets – have left their living space and, if possible, their settlements, to perform their bodily evacuation functions. They did not want the outcome anywhere near where they could smell (or see) it, and especially they did not want it in their homes. Earth closets and outhouses, where these drained into cesspools or middens, were invariably placed away from the house or at the extremity of buildings. Places for 'open defecation' were designated at a distance from, or at the extreme edge of, a village or hamlet. In some parts of the world, this pattern of behaviour still pertains today and is entrenched in cultural norms and conventions about personal purity, and taught to children during toilet training. As towns grew and became more congested in Europe, the problem of excretion and dealing with the result became more acute, and 'houses of office' were forced inside, usually in the basement, often with such unpleasant results that many people continued to use 'close stools' and throw the contents in the road. Only with the welcome invention of the flushing toilet and its widespread take-up as a household improvement did the excretory function come fully indoors.

At that point, toilet installation in homes took off to such an extent that they created the Great Stink and similar environmental crises. In today's world, unless the kind of sanitation offered is sufficiently stink- and noisomeness-free, the millions of people who are without facilities will not want to install them inside their homes, or in their compound if it is relatively small. They will continue to prefer, quite understandably, to go and do their business 'outside'; or, if there is no such possibility, and no 'improved' facility to use in comfort and privacy, they would mostly rather 'wrap and throw'. The most fundamental characteristic of the toilet, that it should be congenial to use on a continuing basis, and should not fill the house with bad odours, is the critical feature of the 'home improvement' it represents. This is much more important from the point of view of users than that it confine excreta where it cannot pose a direct or indirect threat to their health. If this is not understood, a programme to promote household sanitation to bring an end to indiscriminate defecation in banana groves, rice fields or other areas of vegetation will ultimately fail. This ought to be common sense, and some studies do point out, buried away in a section on 'design improvement', that the toilet models currently on offer suffer from 'poor smell'.[12] But some practitioners, who don't have to live with the facilities whose health benefits they recommend, seem to be nasally myopic on this point.

If standards of cleanliness and personal respectability in any given culture have been tied to undertaking excretory functions at a distance from the home where they cause least unpleasantness and embarrassment to everybody concerned, great care has to be taken in proposing the antithesis. Where excreta has routinely been used in agriculture, reservations about handling it or keeping it in an accessible chamber or vault will not be as deeply felt. But still, as the Chinese experience with ecological sanitation in Yongning County demonstrated, even where raw faeces are not treated with superstitious horror but regarded as a fertilizing boon, to construct an indoor toilet and bathroom is a radical departure in domestic lifestyle, and one that would not be considered if the toilet was not smell-free.[13]

By contrast, where a toilet is congenial and effectively deals with odours, it may turn out to be acceptable even in places where people have traditionally had a cultural objection to keeping excreta on their property, in their house or compound, in an above-ground chamber or a pit beneath their feet. A polished ceramic pan that is easy to clean and a decent pour-flush with a water-seal do much to overcome people's resistance to 'going' indoors. The lavatorial experience in peri-urban Dakar or Toamasina, Madagascar, was so transformed by the pour-flush water-seal as to change behaviour overnight. So whatever other quality aspects are encapsulated by words such as 'basic' and 'improved', the question of smell should not be ducked. To meet such a requirement may mean extra expenditure. Cost reductions are attractive, but should not be taken too far. A shallow pit toilet without a vent, a water-seal, a tight-fitting cover on squat-hole or pedestal, or the use of an alternative neutralizer such as ash or lime, cannot reasonably be expected to be sustainably fragrant.

Another lesson that has forced itself to the surface is the need to make a clear distinction between sanitary approaches for urban areas of the developing world and those for rural areas. As we discovered in Chapter 2, the invisibility of many squatter and shanty-town residents in surveys and censuses means that their lack of facilities does not show up in the data collected at the international level.[14] As a result, the relatively rosier picture of sanitation access in urban areas as compared to rural areas – 80 per cent compared to 39 per cent according to the latest computation[15] – ignores those who need services most urgently. The growing numbers of people living in confined spaces in miserable and undignified conditions in the rapidly urbanizing developing world are a cause for major sanitary concern, including threats of cholera epidemics of the kind that plagued 19th-century Europe. Long-standing rural bias in anti-poverty programmes has reinforced neglect of poor urban populations, with regard not only to sanitation, but to water and other basic services.

When density of settlement and housing conditions are put squarely in the framework of lavatorial analysis, the demand for sanitation among poor urban dwellers and the public health case for providing it are unquestionably stronger than for much less densely settled rural areas. And the type of sanitation that can answer the needs of the urban poor, both the household or multi-household fixture and the excreta disposal system, is going to be different for those squashed into tiny living spaces than for those whose compounds are spacious and for whom land tenure and rents are not such major issues. The conventional sewerage solution which has long been seen as *the* industrialized urban excreta disposal solution is not appropriate, and even though small-bore sewerage is growing in importance for the *barrios en desarrollo* of Tegucigalpa and similar settings, its application remains limited. Meanwhile the low-cost 'on-site' solutions with which we have been concerned throughout this book also have important shortcomings for overcrowded *bustis* and townships. These include space, the difficulty of maintaining standards of personal and environmental cleanliness, the stronger stink of effluvia, and the need to get shit out of the house by any means possible – preferably by removal from bucket or pit to a plant where it can be treated, deodorized and rendered harmless, not by 'wrap and throw' or dumping it in the gutter in the middle of the night.

One approach that some practitioners are pursuing is better provision of public toilet blocks, following the model of the Sulabh enterprise in the Indian sub-continent. Because the state of public and multi-use toilets in developing country towns and cities is usually lamentable, at present the definition of 'improved' sanitation used by WHO and UNICEF does not include shared facilities.[16] This may need to be reviewed, and criteria reconsidered: Sulabh facilities are certainly 'improved', and in many settings – even in some very poor and crowded rural settings – shared toilets are the only way forward. In line with this thinking, the UNDP/World Bank Water and Sanitation Program (WSP) in Africa has begun to look into the upgrading of existing public facilities in some locations, notably in Nairobi.[17] Based on a study into the refurbishment, management and operation of public toilets in Ghana, Burkino Faso and Uganda, a five-year pilot plan for repairing and transferring into private hands the management of some of the public toilet blocks owned by the Nairobi City Council was developed and put into action. Here is another area where small-scale providers may find it possible to make a decent sanitary living, under contract to the municipal councils and under their supervision. Certainly, the need for multi-use and shared facilities in urban areas – around markets, shopping malls, bars and food kiosks, in areas where petty entrepreneurs make a living on the streets, as well as in crowded slums and settlements – suggests that organized efforts should be made to bring the public toilet out of its current state of condemnation, by destruction and total rebuilding (as in the case of India's dry latrines) wherever necessary.

When it comes to the private facility, which people with a limited but reasonable amount of living space can aspire to, then the parameters of on-site pit sanitation need to match properly their requirements, purses and habits. In the early days of urbanization in the post-colonial era, families often used their own labour to dig pits deep enough to last the household for a very long time: three metres was not uncommon, and even 'long drops' up to eight or ten metres deep were known in some parts of East Africa.[18] At a depth of around four metres or more, or if a pit is of a very large size (for example for a beer-hall or school), it may never need to be emptied. But households with pits of such a size or depth are becoming less common: they are not usually dug so deep in sanitation programmes today. As living space shrinks, single permanent pits become the only option for many households who want a place of their own. And then regular emptying becomes essential, as we saw in Chapter 6. Up to now, with rural environments more often in mind, the assumption has too often been that the technological problems relating to low-cost sanitation can intrinsically be solved via the initial design.

In the interests of increasing supply and product options, especially those that would permit operation and maintenance for facilities in poor urban areas, hardware issues need to be revisited. Trials with pumps for emptying pits, notably the Vacutug and 'gulper' we met in Chapter 6, and with more conventional truck-mounted and tractor-trailed vacuum tankers, have brought to light features of sanitary 'access' and 'use' that have implications for pit-toilet design. When the fee for the emptying service is based on the number of loads or the volume of waste removed, a poor householder usually wants only enough waste to be removed to solve their immediate problem. The operators therefore suck out through the squatting-hole one load of the liquid waste (around one cubic metre) from the top layer in the pit. The sludge at the bottom continues to solidify and gradually silts up. Thus existing pit-emptying systems are progressively converting deep and large-capacity pits to smaller-capacity pits, and predominantly removing water rather than sludge. This is inefficient and costly to the householder.[19]

In the 1980s, when new handpumps for low-cost community water supplies were being developed, the engineers reached a similar crossroads. Initially, the design challenge was the frequent breakdown of old-fashioned pumps. So O&M problems were addressed by developing new pumps (the India Mark II is the key example) that were cheap and sturdy enough not to break down. But all equipment breaks down at some time, so O&M services were needed nonetheless. At this point, the idea of community maintenance and repair was 'discovered' – it was known as village-level operation and maintenance or VLOM, and a cadre of workers – handpump caretakers – was invented to perform an equivalent function to today's pit-emptying frogmen. Because a VLOM caretaker could not mend an India Mark II handpump as then designed, the next technological hurdle was to design a pump that could be removed from the borehole without heavy lifting gear and had all the features necessary for VLOM. In the case of pit toilets today, the requirements for servicing demand that design parameters evolve, just as for the handpump. A balance is needed between appeal, durability, expense and technically simple O&M.

The question of what is to happen to the sludge is also important. At the moment, much of the excrement from unsewered and unmanaged environments finds it way untreated into rivers, which the development of VLOM emptying services could and should avoid. There are only a limited number of urban environments where sludge can realistically be collected for use in agriculture. But it can also be used, and in some settings is being used, to generate biogas on a commercial scale. Some studies have also been undertaken into the use of 'bio-additives' in pits as a way of accelerating the anaerobic digestion and degradation of solid wastes and reducing the quantity of sludge, thus extending the life of the pit.[20] Certainly there are whole new areas of technological challenge concerning the breaking down of organic outputs from human waste which this book has no space to explore. The biological understanding of the behaviour of microbes, enzymes and nutrients and how they can be used to facilitate the anaesthetization of shit has moved on considerably since the days of the Reverend Moule and 'dry conservancy'.

Let us return to the humble item of domestic use. What changes are needed to the pit toilet itself to allow it to gain a respectable equivalent to the handpump caretaker? According to Manus Coffey, the engineer who designed the Vacutug, the waste from a pour-flush toilet is smaller in volume than that in a VIP type dry toilet, because the organic material decomposes faster and some of the faecal material is dragged away when the water leaches out of the pit. It is also easier to pump out because it is more fluid. Most people who want a water-seal toilet also want a shower to complete their bathroom. Where water supplies are scarce, Coffey's solution is that people save their greywater from laundry and showering, and use it for the flush (Figure 7.2). The size of the pit would be smaller than that needed for dry systems, but would require regular emptying. Coffey has other suggestions for pit design modifications to make emptying easier. These include the idea of a pre-cast pit-tank using stronger but thinner concrete, costing less than conventional lining rings and with better defences against flooding and overflow.

Other technological refinements to all kinds of on-site toilets are under development, as noted in Chapter 4; details can be found in manuals put out by practitioners and research institutions.[21] The example of research and development around the issue of pit-emptying cited above is given here because it addresses critical demands of consumers, and incorporates durability and ease of O&M in the approach. It also illustrates that even so modest a technology as that of the pit toilet requires constant evolution to meet new requirements. And there are many other technical issues that need regular reassessment. One of these is the threat of groundwater contamination by the installation of pits in certain areas.[22] In high water-table areas approaching 'total sanitation' in Bangladesh and West Bengal, for example, the numbers of pits full of raw excreta

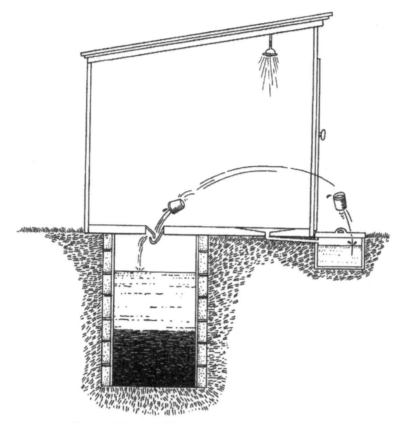

Figure 7.2 *A combined toilet and washroom, Angola*

The shower is situated in the same cabin as the pour-flush toilet, and drains to a small soak-pit. The toilet user carries a bucket and fills it from the soakpit before entering the cabin, using the greywater for flushing.

Source: Manus Coffey (2007)

just below the ground are rising into the millions. Saturation coverage with pit toilets in an area vulnerable to flooding has never been attempted before in such an intensive way.[23] Rules govern the distance to be maintained between pits and open ponds used for bathing and washing and dug wells used for drinking. But in the enthusiasm for 'total sanitation', are they rigorously observed? Regular emptying services may be needed in densely settled 'total sanitation' rural areas in the future.

As the lavatorial scene evolves, a sharper distinction also needs to be made between the spread of toilets and the spread of sanitation systems, which are currently often conflated. As observed in Chapter 6, the on-site facility combines in its features both the role of the toilet and that of the sewer. The toilet – its pan, lid, water-seal and flush, and a deodorizing mechanism such as a vent-pipe or bag of lime – is the personal 'convenience'; the pit is the facility for confining excreta hygienically and preventing it from lying about in the open or being washed into waterways. Thus pits remove, store, and decompose or neutralize the offending matter, just like a sewer and treatment plant. The storage and decomposition part of on-site systems needs to be viewed in the same light as sewerage since it performs the same public health protection function. Water and sanitation utilities should play a similar role to the one played in areas that are sewered, providing the infrastructure for removal, treatment and safe disposal of sludge. Such an approach might mean that the costs of installing mass sanitation would be higher than today's lowest-cost examples: the design would have to be emptying-friendly and the quality of the construction higher, and in order to attain the public health outcomes, a system of regulation and standard-setting would need to be in place. But since the system would involve follow-up services, not all of the costs of 'sustainability' would have to be borne by the household. The manufacture and sale of low-cost toilet components, certified construction of pits and toilet houses, introduction of home improvement loans, and use of public funds to back the system would help promote the mini muck economy on which an urban sanitary revolution could be built.

At the other end of the housing density scale, in parts of the countryside where homes and compounds are scattered, different approaches are needed. Few studies have been conducted into traditional systems of excreta disposal and their implications for public health. But in hot and arid desert zones and sparsely settled mountain areas where livelihoods are close to subsistence, it may be advisable to engage in a dialogue about personal cleanliness, modesty values and safe excreta management before proceeding to the promotion of toilet construction. In places where controlled defecation (by the cat method or faecal burial) in designated places away from habitations is actively preferred, and could well pose less of a health hazard than substitution with a low-quality in-house facility, this may be the most appropriate and viable sanitation system for the moment. In the Himalayan foothills, in the great African deserts, in remote tropical wetlands in Latin America and East Asia, imparting information about excreta-related disease and how to avoid it by washing hands and faces, protecting drinking water, and disposing carefully of faeces – especially those of infants and toddlers – may be as far up the sanitation ladder as it is currently practicable to go. In low-income rural environments generally, the main message ought to be not 'build yourself a toilet', but 'zero tolerance of faecal matter' in any public or open space, including the household compound. From the perspective of those living at or near subsistence, the latter is cost-free and manageable with the use of a small spade; the former may have to wait.

So much for a brief foray into the scientific and technological side of the new sanitary revolution. The much more difficult side – as Edwin Chadwick would echo in bitter frustration – is the creation of the necessary legal, policy and administrative framework, and the associated task of convincing the authorities to allocate the necessary public funds. And however successfully today's legatees of Thomas Crapper and the Reverend Moule manage to market and sell different types of toilet bowl, squat plate or on-site tank, public funds on a significant scale will be required. The story of how low-cost sanitation vanished during the privatization of the water utilities in many developing countries, which we have explored in earlier chapters, is enough to make the point.

This book has assiduously kept to its brief of telling the excretory story and not been diverted except when absolutely necessary onto issues surrounding safe water for drinking, or water for washing, bathing, laundry, cooking and other domestic use. Most enthusiasts for public health protest that sanitation is impossible without water, and that, within the programme, service or management context, the two cannot be separated. In most people's experience they accompany one another in intimate fashion, in the private bathroom and in personal hygiene behaviour. Some public health promoters may be critical that handwashing – sometimes talked up even more than toilets as the key to reducing diarrhoeal disease – has not been given its proper place in these pages, and that 'sanitation' is a much broader concept than human waste disposal and ought to spread its net to include domestic water provision – without which effective human hygiene and 'toilette' is impossible.

The problem is that whenever sanitation is bracketed with water, it becomes the quintessentially poor relation or, as one expert put it during research for this book, 'the ornamental word'. An examination of standard literature on water and sanitation – from the magisterial reports produced by the international system to material from most of the NGOs working in the sector – shows how often this is the case. The words 'and sanitation' are frequently added to 'water' in policy and programme descriptors, without any indication of the need for radically different approaches, technologies, financing methods or mobilization campaigns where sanitation is concerned. Profiles of those affected by excreta-related disease and of efforts on their behalf talk about dirty or unsafe water, but they rarely mention the much more important threat from the pathogens in faeces. Photographs show children drinking from filthy ponds, but the viewer has to use their imagination to understand the true nature of the hazard. Even organizations that could take the lead in tackling the Great Distaste remain linguistically, and conceptually, squeamish about the true nature of the crisis.

This squeamishness confounds the question of how to design the necessary administrative, institutional and managerial infrastructure for sanitation, and get the necessary resources dedicated to it. Many issues or groups which suffer neglect – HIV/AIDS and 'gender', children and the disabled, for example – have to negotiate a balance between being singled out or being 'mainstreamed'. If they remain tucked away in some ministerial or departmental home as a subset of something else, they receive bottom-of-the-list attention and are starved of funds. This is what has happened down the years to sanitation. But if an issue or group is singled out for its own special ministry or department, especially if it is unpopular and subject to discrimination, this risks that the department or ministry in question will be treated as a ghetto. Funds allocated are too few, the least able or energetic civil servants are sent there to work, creating the new bureaucracy becomes an end in itself, and nothing much gets done on the ground. Meanwhile other departments that used to be involved divest themselves of old responsibilities.

So when it comes to issues of governance, should sanitation be pooled with water, or should it stand alone? There is no easy answer. Most programmes integrate water, sanitation and hygiene, at least nominally, and some experts argue that extra interest in water in recent years has released extra funds and that sanitation has 'piggy-backed' on this momentum.[24] It is certainly true that in a world threatened with climate change, floods, droughts, rising urban populations, environmental pressure and possible conflicts over shared freshwater resources, the management of water has become better acknowledged as a vitally important and politically significant concern. But most of the newly won attention has been on water scarcity, irrigation, dams, energy, thirsty cities and flood protection, not on public health. If sanitation stays within the administrative scope of 'water', even when the water concerned is 'water for health' not water for all economic, social and environmental uses, the evidence is that it remains an orphan.

Just to take one example, of the meagre 0.3 per cent of government expenditure dedicated in Madagascar to sanitation and water, 90–95 per cent is spent on drinking water. US$0.005 per person is left for sanitation, a miserable half a cent per year. What can possibly be done with that, laments Rakatondrainibe Herivelo, the government coordinator for WASH?[25] Madagascar is an exceptionally poor country, but in terms of budgetary proportions in the sector its allocation is not unusual. A figure of 1 per cent of national expenditure for domestic water and sanitation is typical (usually without the proportion for sanitation being specified or a special budget being allocated).[26]

The Madagascan WASH review also describes the fragmentation of responsibilities for sanitation within the government; these are distributed between the Ministries of Land and Urban Planning, Health, Energy and Mines, Environment, and Industry and Handicrafts, plus the agency responsible for devolution and decentralization of community governance. In almost every country, the pattern is similar, often because responsibilities for water are also fragmented and sanitation merely follows in its wake. Coordination between all these bodies is a nightmare even at the central level; once the lower layers are also brought into the picture – regional, district, municipal, village, community – the result is often a tangled web of overlapping, uncoordinated, unworkable policies and programmes with low budget allocations and little prioritization.[27] Making progress in such an environment requires leaders of determination, conviction and outspokenness – leaders who are few and far between, and many of whom prefer to focus on more salubrious parts of their brief.

This book has so far been determined to de-link sanitation from water – a coupling that has led to so much misdiagnosis of sanitary problems, so many inappropriate responses, and so much ignorance and misguided assumptions about what people really want in the way of human-waste disposal systems. Accordingly, it seems appropriate also to support some separation of sanitation from water supply authorities, with the understanding that the two also have to work hand in hand, and that sanitation and hygiene have as valid a claim on water supplies as agriculture, energy, health and environmental conservation, demanding consideration alongside drinking water supplies. Some actors in the sector – for example the Dutch NGO WASTE – are beginning to develop partnerships with municipal utilities in charge of solid waste disposal, finding that they have more of a common agenda around collection and safe disposal of sludge than they do with authorities in charge of public health.[28] Voices that today demand for sanitation its own institutional 'home' should be heard, even while continuing to involve other sectoral partners and keeping the overall approach multidisciplinary.

Whether this 'home' should consist of a separate ministry or national department, or be carved out of, but remain within, something else – lands and/or water, municipal affairs and/or local government, public and/or environmental health – will need to be considered on a country-by-country basis. The purpose

of any reform in institutional management would be to put resources and quali-
fied staff in place and provide training, capacity and support to sanitation and
hygiene officers at every level from the community upwards. There has to be a
presence, a committed and vocal sanitation leader who understands all the parameters, at the top governmental table. Then clout can be exercised and real
momentum and change may start to occur. Many of the changes will have to
take place not at the centre, but at the level of district and local government and
community affairs. Top–down efforts must be used to promote bottom–up
consultation and bring about the spread of services that generate and satisfy
genuine demand.

At the international level, the UN Year of Sanitation offers an opportunity
to exert pressure on national governments to undertake institutional reforms.
But the primary opportunity it offers is to raise the profile of sanitation throughout the international donor and NGO community, and bring to professional and
public consciousness both the scale of the sanitary crisis and its true nature. The
first priority is to make sure that analyses of poverty at both the national and
international levels in which donors cooperate do not overlook sanitation
altogether. The much-touted tool for this in recent years has been the poverty
reduction strategy paper (PRSP), a joint effort by national governments and
donor organizations, led by the World Bank and International Monetary Fund,
to develop a targeted plan of inputs and outputs to reduce poverty across a
concerted front in any given country. Water and sanitation, however, and
especially the latter, make rare appearances in these PRSPs (except in the context
of water utility privatization). There is a lack of appreciation that poverty and
lack of sanitation and clean drinking water are closely interrelated, and that access
to basic services is a key aspect of poverty reduction.

What of financing? As WaterAid has pointed out, because few countries
have national sanitation policies, and since it does not feature in development
plans, sanitation is not considered for funding by many donors.[29] Developing
country governments themselves are partially to blame. In some countries,
especially in sub-Saharan Africa, there is still barely any recognition that sanitation belongs in the public domain – it is seen as a purely private matter. The
Global Water Partnership estimated in 2000 that, while US$13 billion per year
was spent by donors on water, just US$1 billion was committed to sanitation.[30]
More recent figures from the Joint Monitoring Programme of WHO and
UNICEF suggest a proportion of eight to one, but this is just as depressing
since these figures relate only to domestic water supplies, not water for agricultural and other purposes. In cases where donors have been active on the

sanitation front, they have often gone in for large and unsustainable urban infra-structure projects financed by loans, which do little or nothing to improve matters for the poor.[31]

Donor contributions need to be upwardly revised – by leaps and bounds. However, to put a figure on the necessary amount would not be helpful. These kinds of calculations are undertaken as part of the MDGs exercise, as they earlier were for the Water and Sanitation Decade. Given that the kind of programmes needed are quite unlike the single agency, capital intensive, 'lumpy' investments of traditional public health engineering, but depend on multiple small invest-ments dovetailed to local situations over long periods, synthesized estimates of global costs are inevitably of the virtual variety. However, figures computed in billions of dollars for meeting the sanitation target have been produced – for example, WHO has suggested US$9.5 billion annually over and above current investments, and there are others.[32]

Not only are such figures discouraging, because it is difficult to conceive that resources of this magnitude could suddenly be found for sanitation programmes in competition with so many other more popular targets, they also erroneously imply that the chief gaps are financial, and that the Goals could be realized if only sufficient quantities of aid and investment were pushed in their direction. But if the necessary policies, political will and institutional capacity for spending money well are not in place – tasks that themselves require investment – the outcomes of programmes to build and satisfy demand for toilets and sanitation systems will be less effective. It takes time and resources to develop the neces-sary policy and administrative frameworks. Without basic underpinning, the many activities we have already identified – piloting different models and systems, training outreach workers, consulting with communities, finding ways to work with and capitalize local entrepreneurs – cannot be carried out effectively, except by NGOs on a small, scatter-gun and serendipitous scale. Where the develop-ment of governmental infrastructure is concerned, there are no short cuts, and the more sensitive the issue, the longer it takes.

Existing failures to address the fundamental human need for decent sanita-tion reflect the unwillingness in societies everywhere to talk about excreta disposal and behave as if it was a matter of public importance instead of private embarrassment and shame. The related and false assumption that there is no demand among the poorer inhabitants of the planet for places to perform their bodily functions in a safe and dignified way and have the outcome hygienically removed helps to explain, but not to excuse, its absence from policy debate. Political leadership is sorely needed to champion the new sanitary revolution and turn this situation around. In Chapter 5 we met the 'latrine *Bupati*' of Lombok, Indonesia, whose 'Clean Friday' movement led to Presidential recognition and involvement. We have also encountered Bindeshwar Pathak of India, Rafael Díaz of Nicaragua, Edwin Joseph of Madagascar and many other individuals in government offices, health departments, NGOs and communities who have pioneered a breakthrough in lavatorial awareness and environmental change. But these examples are not nearly so numerous as they could and should be.

People in leadership positions high and low need to be confronted with the true parameters of the sanitary crisis so as to encourage them into taking up the cause. Perhaps, like the President of Indonesia, they can champion the cause of 'clean and pure' communities, instead of being expected to talk of unmentionable dirt and shit.

One person who has made a start is Sim Jae-Duck of Korea, chairman of a new organization, the World Toilet Association. He has built a huge steel, glass and concrete house in the shape of a toilet (see Figure 7.3), called *Haewoojae* (signifying 'a place of sanctuary where one can solve one's worries' in Korean). The house is part of his effort to foster an open dialogue about toilets and eradicate taboos and misconceptions. Sim is today's most exuberant toilet missionary,

Figure 7.3 *Sim Jae-Duck's* Haewoojae *'toilet house' in Korea*

This image is a computer-generated design for the house, construction of which was completed in late 2007.

Source: World Toilet Association, http://en.wtaa.or.kr

and is raising money to improve facilities and hygiene standards around the world. His methods seem odd, but perhaps they will confer a touch of glamour and bring others on board.

The recruitment of celebrity names would add momentum to national mobilization efforts and annual 'sanitation' or 'cleanliness' drives. Presidents, pop stars and football heroes should be persuaded to bracket their names to a campaign against the Great Distaste, as they already do to the more attractive and popular issue of 'clean water'. At the community level, opportunities for self-expression must be given to those who feel distress about their lack of sanitary facilities but are currently inhibited from giving it voice. Every effort must be made to overcome timidity and fear of being thought vulgar or socially impure, so that the subject can finally be brought out of the closet and into the air.

B efore we wrap up our sanitary excursion, it is worth re-emphasizing one of the most important features of today's sanitary crisis: that this is essentially a crisis affecting the poor. It is no coincidence that the numbers of people estimated to be without access to a decent toilet facility – 2.6 billion – are almost equivalent to the 2.5 billion estimated to be living on less than US$2 a day.[33] In surveys that dissect coverage figures and find out exactly who has what in poor communities, it is invariably the poorest households that have the fewest and worst toilet facilities (Figure 7.4). They also have the hardest time finding the cash resources to install something better – or anything at all.

Among those without access to basic sanitation, those who suffer worst in terms of sanitation-related illness are children under five, among whom diarrhoeal diseases remain one of the biggest killers worldwide; and those who suffer worst from discomfort, indignity and insecurity are women, adolescent girls, the elderly, the disabled, the sick and the infirm. This applies both in their households and in places which are used communally: schools, clubs, health centres and hospitals; homes for old people; homes for those with disabilities or orphaned children; and penal institutions where children, youths and women are held. The indignities that these people on the very bottom rung of societies with deficient service infrastructure have to endure to perform their bodily functions have to be seen to be believed.

In the 19th and early 20th centuries, the public health revolution precipi-tated by Chadwick and the other sanitary heroes was eventually extended – via local government financing, municipal sewerage, housing regulations and the

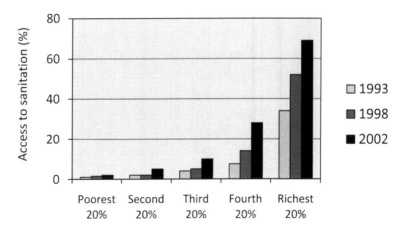

Figure 7.4 *In Vietnam, the rural poor are left behind*

Source: K. T. Phan, J. Frias and D. Salter (2004) *Lessons from Market-Based Approaches to Improved Hygiene for the Rural Poor in Developing Countries*, 30th WEDC International Conference, Vientiane, Laos

marketed promotion of lifestyle change – to the entire populations of industrialized countries. Sewerage exerted a democratizing force, as could the sanitary revolution of today. In earlier times, the threat of infectious disease in the urban environment, along with greater prosperity, helped impel the revolution forward. In the modern era, there has not been the same impetus from life-threatening epidemics to promote the spread of services to all levels of society, partly because nowadays we have the medical means to tackle diarrhoeal and other kinds of epidemic disease via health service interventions. Greater prosperity and changing lifestyles are making their mark, even if municipal reform, housing improvements and financing still languish far behind. What will take the place of the impetus of life-threatening disease? High pollution levels in rivers, from which everyone suffers, seem to be one of the few cards left in play.

In Tegucigalpa, Honduras, the foul smell of the River Choluteca, flowing through the centre of the town and during the dry summer months containing virtually nothing but sewage, was a source of disgust to every citizen (see Chapter 2). Eventually, this led to programmes for simplified sewerage in those parts of town – the poorer parts – without any means of sanitation, and to mass sewage treatment. In India, the story of efforts to clean up the River Yamuna has a similar derivation, but not yet as positive an outcome. The Yamuna flows through Delhi, collecting the city's excrement, and its putrefying condition led to the construction of 15 new sewage treatment plants to collect and process the content of Delhi's sewers so as to clean the river up.[34] But an equivalent effort has not been put into appropriate sanitation for those living in the slums. When Sunita Narain of the Centre for Science and Environment went to visit the plants, she discovered that once the effluent had been expensively treated, it was released into stormwater drains leading back to the river. These drains carry the excreta from adjoining squatter colonies, in which no kind of sanitation has been introduced. So the content of the Yamuna remains disgusting and will continue to do so while sanitation for poor communities is ignored. Perhaps it will take repeated Great Stinks from which everyone suffers, as well as the death of rivers now carrying a load of pollution they cannot absorb, to finally build political demand behind sanitation in many cities and countries, not just for those who can afford sewer connections, but also for those who can't.

Equity in sanitation is not only a question of public action on behalf of sanitation for the poor. It also requires that programmes targeted at low-income areas, urban and rural, do not demand full cost recovery from new users. Perhaps we should revert to the formula adopted in the Victorian age, in which the system of removal – whether by sewer and treatment or pail collection and storage – is vested in the authorities, and the purchase of the toilet pan is left to the household. But this still leaves the pit, its lining if it needs one, pipes and vents leading in or out, and above ground vaults and their accoutrements in an either/or position – they are not exactly 'pan' or 'slab', and not exactly sewer. Layered and means-related subsidies to cover these elements, as operated in Toamasina, Madagascar (see Chapter 6), is a good way forward for poverty-stricken areas. Where people are living on the edge of subsistence, they cannot afford unsubsidized construction of all that is comprised above and below ground in a decent toilet house; even with loans, they may not be able to afford the interest and repayments. Important as it may be to avoid subsidizing those who can afford to build and maintain an on-site toilet, the bad record of subsidies in certain settings should not be used as an argument to discount their use altogether. There is no one solution for financing community sanitation. A combination of different approaches is needed. If policies insist on capital cost recovery from very poor households, many of the people who currently survive on less than US$2 a day will continue to be excluded.

What of the especially vulnerable groups within the wider category of those living on less than US$2 a day? Gender advocates have made an impeccable case that women's needs for dignity, privacy and security place them in a special situation. This is reinforced by the expectation that they should invariably deal with the excreta of young children, and that of elderly, sick and disabled people in the household for whom they have to care. Ignorance of women's needs, including the management of menstruation, has dogged many sanitation programmes. Take the case of a typical public toilet block in a South African township. The toilets have been built to face the street, causing women embarrassment and harassment. The cubicles are too small for use by pregnant women and women with small children. No provision has been made for the disposal of sanitary pads. Water for flushing the toilet has to be fetched by women, and it is they who clean the toilet and empty it whenever it is full. Women who perform the latter task are seen as unmarriageable.[35] Here in microcosm are all the ways in which facilities can fail to meet women's sanitary needs and end up by demeaning them and undermining their personal dignity. On the other hand, where women have been invited to contribute their ideas to planning and implementation of sanitation schemes, the results are invariably more successful in gaining local ownership and acceptance.

An even more neglected group in sanitation are the disabled. For those with some kind of impairment that restricts mobility, the presence in the compound or around the corner of an 'improved' toilet facility is not the same as 'access'. As many as one in five of the world's poorest people are disabled and suffer an even greater degree of exclusion from services of all kinds than do their able-bodied peers.[36] Some household sanitation surveys have captured the very special predicament of those with disabilities in relation to sanitation. In Namibia, for example, households with disabled members are less likely to have an in-house flush toilet and more likely to resort to the bush.[37] Children with disabilities are often withheld from school, the lack of decent toilets that they could use being one of the reasons for their exclusion.

Children generally are a critical subgroup for the new sanitary revolution. This needs to start from their very youngest days. Among mothers of small children, hygienic disposal of infants' and toddlers' faeces need to be given more attention in postnatal wards and mother-and-child healthcare programmes. As toddlers grow and gain in understanding, toilet training and the inculcation of good sanitary behaviour both at home and in early childhood centres needs attention – day-care workers should be given relevant instruction, for example. One of the most important international sanitation initiatives currently underway is school sanitation and hygiene education (SSHE) or WASH in schools – programmes not only to provide decent toilet and hand-washing blocks but to inculcate in children a culture of personal cleanliness and toilet use and the desire to continue to apply such ideas back in the home. Where effectively delivered, these programmes have all sorts of useful spin-offs.

For example, in Medak District of Andhra Pradesh, India, little books of soap leaves – like books of matches – with one leaf enough for one wash of hands have been manufactured by local companies and distributed to primary schools. Students line up before meals to receive their tiny soap leaf from a member of staff, who supervises and streamlines the process. These soap books are now part of the everyday equipment of Medak schools. Thus the local private commercial sector has been brought in to help children to adopt sensible lavatorial habits and entrench new behavioural codes. Youngsters who become familiar with the use of soap and toilets at school, and convey to their parents that home life is not wholesome in the absence of a 'necessary room' and a bar of soap next to the hand-washing basin, are tomorrow's consumers of home improvements. Toilet habits should be part of education and 'life skills' in every day-care centre and primary school throughout the world. Institutional facilities in schools and other public settings should set a standard which builds appreciation for comfortable, clean and smell-free excretion and promotes personal hygiene.

Let us sign off by meeting 16-year-old Doly Akter, the leader of an adolescent girls' 'hygiene monitoring group'.[38] This group of girls has not only been converted to toilets at school, but has started to transform the state of the slum where they live in Dhaka, Bangladesh. 'It used to be such a dirty and undeveloped area, but already there are some differences,' says Doly. She and her friends set out on rounds from house to house, asking the neighbours whether they are washing their hands after defecation and before eating, using their toilets and keeping them clean, and drinking only safe and protected drinking water. All these answers are registered by the girls. Over time, their promotion of toilets and hygienic behaviour has reduced the incidence of diarrhoeal disease by half in the neighbourhood. According to Doly, the children in the slum laugh and play more now that they are healthier: 'If we can continue this work, in the future they won't feel any diseases.' Her parents appreciate what she has learned and what she has done, and they too notice the reduction in diarrhoeal attacks. With their confidence boosted, the group has now started to take on another social issue: that of under-age marriages. So sanitation does not necessarily have to be the last taboo; it can, in the hands of Bangladeshi girls, become the entry point for other steps along the path to child rights, gender equity and social justice.

Doly and her friends can do what they are doing because something has intervened to take away the timidity and embarrassment which might otherwise have inhibited their neighbourhood visits. When children come calling to talk openly and seriously about such things as 'zero tolerance of open defecation', their elders – whatever they think – cannot easily brush them aside; they too start to put aside their inhibitions. Despite all the problems surrounding this most difficult of subjects, words can be found and can be used to talk about what needs to happen, by children, by local health or sanitation workers, by engineers, by politicians, by scene-setters and policymakers. The time has come to dispense with the Great Distaste, to mobilize words, efforts and resources, and to do whatever it takes to bring on a new sanitary revolution throughout the developing world.

Notes

Chapter 1

1   UNICEF (2006) *Progress for Children: A Report Card on Water and Sanitation*, UNICEF, New York, www.unicef.org/publications/files/Progress_for_Children_No._5_-_English.pdf.

2   UNDESA (2006) *The Millennium Development Goals Report 2007*, United Nations Inter-Agency and Expert Group on MDG Indicators, New York.

3   Virginia Smith (2007) *Clean: A History of Personal Hygiene and Purity*, Oxford University Press, Oxford.

4   Manof Nadkarni (2002) 'Drowning in human excreta', *Down to Earth*, vol 10, no 19, Centre for Science and Environment, New Delhi.

5   Robin Clarke and Janet King (2004) *The Atlas of Water*, Earthscan, London.

6   Sowmyaa Bharadwaj and Archana Patkar (2004) *Menstrual Hygiene and Management in Developing Countries: Taking Stock*, Junction Social, Mumbai, India.

7   Harold Farnsworth Gray (1940) 'Sewerage in ancient and mediaeval times', *Sewage Works Journal*, vol 12, no 5, pp939–946.

8   Lucinda Lambton (1995) *Temples of Convenience and Chambers of Delight*, Pavilion Books, London.

9   Liu Jiaya and Wang Jungqi (2001) 'The practice, problem, and strategy of ecological sanitary toilets with urine diversion in China', in *First International Conference on Ecological Sanitation, Nanning, China, November 2001, Abstracts*, Jiusan Society and UNICEF, China, and Sida, Sweden.

10 David Eveleigh (2002) *Bogs, Baths and Basins: The Story of Domestic Sanitation*, Sutton Publishing, Stroud, UK.

11 Uno Winblad and Mayling Simpson-Hébert (eds) (2004) *Ecological Sanitation* (revised edition), Stockholm Environment Institute, Stockholm.

12 Jean-François Pinera and Lisa Rudge (2005) 'Water and sanitation assistance for Kabul: A lot for the happy few?', paper submitted to the 31st WEDC International Conference, Kampala, Uganda, http://wedc.lboro.ac.uk/conferences/pdfs/31/Pinera.pdf.

13 Gray, op cit (Note 7).

14 François de la Rochefoucauld (1784) *A Frenchman's Year in Suffolk, 1784*, edited and translated by Norman Scarfe (1988), Suffolk Records Society, vol XXX, UK.

15 Stephen Halliday (1999) *The Great Stink of London*, Sutton Publishing, Stroud, UK.

16 Donald Reid (1991) *Sewers and Sewermen – Realities and Representations*, Harvard University Press, Cambridge, MA.

17 Eveleigh, op cit (Note 10).

18 Halliday, op cit (Note 15).

19 Thomas Cubitt (1840) *Parliamentary Papers*, House of Commons, London, vol 11, q3452: Cubitt's evidence to the Select Committee on the Health of Towns.

20 J. J. Cosgrove (1909) *History of Sanitation*, Standard Sanitary Mfg. Co., Pittsburgh, PA, pp104–106.

21 F. B. Smith (1979) *The People's Health 1830–1910*, Croom Helm, London, pp196–198.

22 Cosgrove, op cit (Note 20), Chapter VII.

23 Gerard T. Koeppel (2000) *Water for Gotham: A History*, Princeton University Press, Princeton, NJ.

24 Eveleigh, op cit (Note 10), p11.

25 Valerie Curtis and Adam Biran (2001) 'Dirt, disgust, and disease: Is hygiene in our genes?', *Perspectives in Biology and Medicine*, vol 44, no 1 (winter 2001), pp17–31.

26 This account is from Cosgrove, op cit (Note 20), who bases his account on Sedgewick, a late 19th-century authority.

27 Christopher Hamlin and Sally Sheard (1998) 'Revolutions in public health: 1848 and 1998?', *BMJ*, vol 317, pp587–591, quoting Smith, op cit (Note 21).

28 Smith, op cit (Note 21).

29 Cosgrove, op cit (Note 20), pp97–99.

30 Anthony S. Wohl (1983) *Endangered Lives: Public Health in Victorian Britain*, J. M. Dent and Sons, London, p65.

31 Smith, op cit (Note 21).

32 Wohl, op cit (Note 30).

33 Wohl, op cit (Note 30), p100, quoting J. Hollingshead (1861) *Ragged London in 1861*, International Health Exhibition 1884.

34 Hamlin and Sheard, op cit (Note 27).

35 Susan E. Chaplin (1999) 'Cities, sewers and poverty: India's politics of sanitation', *Environment and Urbanization*, vol 11, no 1, pp145–158, IIED, London.

36 Martin Daunton 'London's "Great Stink" and Victorian urban planning', www.bbc.co.uk/history/trail/victorian_britain/social_conditions/victorian_urban_planning_03.shtml, accessed July 2006.

37 *The Builder*, 8 September 1861, London, quoted in Eveleigh, op cit (Note 10).

38 Eveleigh, op cit (Note 10), p46.

39 G. Wilson (1873) *A Handbook of Hygiene*, quoted in Eveleigh, op cit (Note 10).

40 Eveleigh, op cit (Note 10), pp57–58.

41 Wohl, op cit (Note 30), p 79, quoting B. S. Rowntree (1901) *Poverty: A Study of Town Life*, p214.

42 Wohl, op cit (Note 30), p101, quoting Stevenson in *Bulletin of the History of Medicine*, vol XXIX, no I (January/February 1955) p11.

43 Wohl, op cit (Note 30), pp98–99.

44 Wohl, op cit (Note 30), p101, quoting Parliamentary Paper XXXVIII (1876) *Report of a Committee Appointed by the President of the Local Government Board to Inquire into Several Modes of Treating Town Sewage*, ppxii and xiii.

45 Wohl, op cit (Note 30), p107, quoting C. B. Chapman (1972) 'The year of the Great Stink', *The Pharos*, vol 35, p94.

46 F. B. Smith, op cit (Note 21).

47 F. B. Smith, op cit (Note 21), p201.

48 M. W. Flinn (1968) *Public Health Reform in Britain*, MacMillan, London.

49 Ibid.

50 B. R. Mitchell and Phyllis Deane (1962) *Abstract of British Historical Statistics*, Cambridge University Press, Cambridge, pp34–43; quoted in F. B. Smith, op cit (Note 21).

51 F. B. Smith, op cit (Note 21).

52 Flinn, op cit (Note 48).

53 Leonard Metcalf and Harrison P. Eddy (1914) 'The lessons taught by early sewerage works', in Barbara Gutmann Rosencrantz (1977) *Sewering the Cities*, Arno Press, New York; reprinted from *American Sewerage Practice*, vol 1, New York.

Chapter 2

1 Mike Davis (2006) *Planet of Slums*, Verso, London.

2 Christopher Flavin (2007) 'Preface', in *State of the World 2007: Our Urban Future*, The Worldwatch Institute, Washington, DC, and W. W. Norton, New York, pxxiii.

3 Deepa Joshi (2005) 'Sanitation for the urban poor: Whose choice, theirs or ours?', report of a DFID-funded research project, managed by the Institute of Irrigation and Development Studies, University of Southampton, pp51–53; calculation based on the 13 million houseless households recorded in the 2001 census.

4 Kai N. Lee (2007) 'An urbanizing world', Chapter 1 in *State of the World 2007: Our Urban Future*, The Worldwatch Institute, Washington, DC, and W. W. Norton, New York.

5 Davis, op cit (Note 1), p20, citing Anqing Shi (2000) *How Access to Urban Potable Water and Sewerage Connections Affects Child Mortality*, Finance Development Research Group, The World Bank, Washington, DC.

6 Janice E. Perlman and Molly O'Meara Sheehan (2007) 'Fighting poverty and environmental injustice in cities', Chapter 9 in *State of the World 2007: Our Urban Future*, The Worldwatch Institute, Washington, DC, and W. W. Norton, New York.

7 WaterAid and Tearfund (2003) *New Rules, New Roles: Does PSP Benefit the Poor?*, WaterAid and Tearfund, London, p33.

8 UN Habitat (2003) *Water and Sanitation in the World's Cities*, UN Habitat, Nairobi, and Earthscan, London, p3.

9 Maggie Black (2007) author's personal experience, field visit organized by UNICEF, Dakar.

10 Joshi, op cit (Note 3), p101.

11 Human Rights Watch (2005) *Zimbabwe: Evicted and Forsaken – Internally Displaced Persons in the Aftermath of Operation Murambatsvina*, Human Rights Watch, New York, http://hrw.org/english/docs/2005/12/01/zimbab12111.htm.

12 Martin Mulenga, Gift Manase and Ben Fawcett (2004) *Building Links for Improved Sanitation in Poor Urban Settlements*, Institute of Irrigation and Development Studies, University of Southampton, Southampton, UK, p16.

13 Maggie Black (1995) *Children First: The Story of UNICEF*, Oxford University Press, Oxford, and UNICEF, New York.

14 UNICEF (2000) *Waterfront*, no 14, leading article, p2, UNICEF WES Division, New York, www.unicef.org/wes/files/wf14e.pdf.

15 UNICEF (1998) *Waterfront*, no 12, UNICEF WES Division, New York, www.unicef.org/wes/files/wf12e.pdf.

16 Mary Racelis and Angela Desirée M. Aguirre (2005) *Making Philippine Cities Child-friendly*, Institute of Philippine Culture, Ateneo de Manila University, Manila, and UNICEF ICDC, Florence, Italy.

17 Perlman and Sheehan, op cit (Note 6).

18 Joshi, op cit (Note 3), p49.

19 Davis, op cit (Note 1), p72.

20 Brian Appleton (1988) *Towards the Targets: An Overview of Progress in the First Five Years of the International Drinking Water Supply and Sanitation Decade*, WHO, Geneva.

21 WHO and UNICEF (2006) *Meeting the MDG Drinking Water and Sanitation Target: The Urban and Rural Challenge of the Decade*, WHO/UNICEF Joint Monitoring Programme, Geneva and New York, www.wssinfo.org/pdf/JMP_06.pdf.

22 David Satterthwaite and Gordon McGranahan (2007) 'Providing clean water and sanitation', in *State of the World 2007: Our Urban Future*, The Worldwatch Institute, Washington, DC, and W. W. Norton, New York.

23 WHO and UNICEF, op cit (Note 21), Figure 18, p19.

24 Mulenga, Manase and Fawcett, op cit (Note 12), p24.

25 Maggie Black (2006) visit to Nicaragua, meeting of the Sector-Wide Approach Group for the Water and Sanitation Sector, Working Group Presentation on Integral Sanitation, October 2006.

26 Joshi, op cit (Note 3).

27 UN Habitat, op cit (Note 8), pp86–90.

28 Meera Bapat and Indu Agarwal (2003) 'Our needs, our priorities; women and men from the slums of Mumbai and Pune talk about their needs for water and sanitation', *Environment and Urbanization*, vol 15, no 2, pp71–86, IIED, London.

29 WHO and UNICEF, op cit (Note 21), India sanitation data from various data sources.

30 UN Habitat, op cit (Note 8), pp16–18 and 23–26.

31 Satterthwaite and McGranahan, op cit (Note 22).

32 UNDP (2006) *Beyond Scarcity: Power, Poverty and the Global Water Crisis*, UN Human Development Report 2006, New York, p112.

33 WHO and UNICEF, op cit (Note 21), p21.

34 Marion W. Jenkins and Steven Sugden (2006) 'Rethinking sanitation: Lessons and innovation for sustainability and success in the new millennium', Occasional Paper 2006/27, UN Human Development Report, New York.

35 Satterthwaite and McGranahan, op cit (Note 22), p35.

36 Madeleen Wegelin-Schuringa (2000) 'Strategic elements in water supply and sanitation services in low-income urban areas', *Waterfront*, no 14, UNICEF WES Division, New York, pp3–6.

37 WSP (2001) 'El Alto Condominial Pilot Project impact assessment: A summary', WSP field note, Water and Sanitation Program Latin America, Lima.

38 Joshi, op cit (Note 3), p60 passim.

39 UNDP, op cit (Note 32), p124.

40 UN Habitat, op cit (Note 8), p232.

41 David Satterthwaite (2006) 'Appropriate sanitation technologies for addressing deficiencies in provision in low- and middle-income nations', Occasional Paper 2006/30, UN Human Development Report, New York.

42 UNDP, op cit (Note 32), p126.

43 Maggie Black (2006) author's personal information, visit to Tegucigalpa and meetings at UEBD and at project sites, October 2006.

44 UN Habitat, op cit (Note 8), p176.

45 UNDP, op cit (Note 32), p114.

46 CSE (2007) *Sewage Canal: How to Clean the Yamuna*, Centre for Science and Environment, New Delhi.

47 WSP (2003) 'Urban sewerage and sanitation: Lessons learned from case studies in the Philippines', WSP field note, The World Bank, Washington, DC.

48 Maggie Black (2007) author's personal experience, visit to AGETIP programme in urban periphery of Dakar, Senegal, March 2007.

49 WaterAid and Tearfund, op cit (Note 7), case study from Tanzania, p34.

50 Mulenga, Manase and Fawcett,op cit (Note 12), p22.

51 Sandy Cairncross (2004) 'The case for marketing sanitation', WSP-Africa field note, The World Bank, Nairobi, Kenya, www.wsp.org/filez/pubs/af_marketing.pdf; cited in Jenkins and Sugden (2006), op cit (Note 34), p4.

52 Mulenga, Manase and Fawcett, op cit (Note 12), p 23.

53 Maria S. Muller (ed) (1997) *The Collection of Household Excreta: The Operation of Services in Urban Low-income Neighbourhoods*, WASTE Environmental Systems Information Center, Gouda, The Netherlands.

54 Fred Pearce (2000) 'It's not just high-flown ideas…', *New Scientist*, vol 166, no 2240, p14.

55 The World Bank (1992) *The World Development Report 1992, Development and the Environment*, The World Bank, Washington, DC.

56 Susan E. Chaplin (1999) 'Cities, sewers and poverty: India's politics of sanitation', *Environment and Urbanization*, vol 11, no 1.

Chapter 3

1 Sandy Cairncross and Richard Feachem (1993) *Environmental Health Engineering in the Tropics – An Introductory Text* (second edition), John Wiley and Sons, London.

2 Sandy Cairncross (2003) 'Water supply and sanitation: Some misconceptions', editorial, *Tropical Medicine and International Health*, vol 8, no 3, pp193–195.

3 Ana Gil, Claudio Lanata, Eckhard Kleinau and Mary Penny (2004) *Children's Faeces Disposal Practices in Developing Countries and Interventions to Prevent Diarrhoeal Diseases – A Literature Review*, Instituto de Investigacion Nutricional (Peru), prepared with the support of the Environmental Health Project, Bureau for Global Health, USAID, Washington, DC.

4 Ibid, quoting studies by R. G. Feachem et al (1978), R. J. Levine et al (1976) and B. Young et al (1988); see also Sandy Cairncross (1989) 'Water supply and sanitation: An agenda for research', paper for the Commission on Health Research for Development, reprinted from *The Journal of Tropical Medicine and Health*, vol 92, pp301–314.

5 See the seminal study by Steven A. Esrey (1996) 'Water, waste and well-being: A multicountry study', *American Journal of Epidemiology*, vol 143, no 6; also more recently, Ricardo Fuentes, Tobias Pfutze and Papa Seck (2006) 'Does access to water and sanitation affect child survival? A five country analysis', Occasional Paper 2006/4, UN Human Development Report, New York.

6 Lorna Fewtrell, Rachel Kaufmann, David Kay, Wayne Enanoria, Laurence Haller and John Colford (2005) 'Water, sanitation and hygiene interventions to reduce diarrhoea in less developed countries: A systematic review and meta-analysis', *Lancet Infectious Diseases*, vol 5, no 1, pp42–52.

7 Barbara Evans (2005) 'Securing sanitation: The compelling case to address the crisis', report commissioned by the government of Norway as an input to the United Nations Commission on Sustainable Development (CSD), published by The Stockholm International Water Institute, Stockholm, p7.

8   E. G. Wagner and J. N. Lanoix (1958) *Excreta Disposal for Rural Areas and Small Communities*, WHO Monograph Series No 39, WHO, Geneva.

9   Sarah Boseley (2007) 'Sanitation rated the greatest medical advance in 150 years', *The Guardian*, 19 January 2007.

10   Evans, op cit (Note 7), p7.

11   Government of Madagascar, WaterAid and UNICEF (2003) *Assainissement: Le Défi*, Government of Madagascar, WaterAid and UNICEF, Antananarivo, Madagascar; and Maggie Black (2007) interview with Jean Herivelo Rakotondrainibe, National Coordinator for Sanitation, Madagascar, March 2007.

12   Richard Feachem, David Bradley, Hemda Garelick and Duncan Mara (1983) *Sanitation and Disease: Health Aspects of Excreta and Wastewater Management*, Johns Hopkins University Press, Baltimore, MD, p9.

13   UNICEF (2005) 'WASH strategy paper 2006–2015', UNICEF, New York.

14   UNICEF (2000) *Sanitation for All: Promoting Dignity and Human Rights*, UNICEF, New York.

15   WHO (2003) *Looking back, Looking Ahead: Five Decades of Challenges and Achievements in Environmental Sanitation and Health*, WHO, Geneva.

16   André Stanghellini (2006) *Rapport de la Mission d'évaluation Finale du Projet SEN/010*, Lux Development SA, Agence Luxembourgeoise pour la Coopération au Développement, Luxembourg.

17   Esrey, op cit (Note 5), p621.

18   Government of Madagascar, WaterAid and UNICEF, op cit (Note 11).

19   BBC (2007) 'Ethiopia cholera death fears grow', http://news.bbc.co.uk/2/hi/africa/6290373.stm, accessed 23 January 2007.

20   BBC (2007) 'Cholera reaches Congo's capital', http://news.bbc.co.uk/2/hi/africa/6331233.stm, accessed 5 February 2007.

21   Government of Madagascar, WaterAid and UNICEF, op cit (Note 11).

22   Andry Ramanantsoa (2004) 'Rapport final:Capitalisation et recherche de solutions sur les latrines à Madagascar', WaterAid Madagascar, Antananarivo, Madagascar.

23   National Water Supply and Environmental Health Programme and WSP (2001) *Consumer's Choice: The Sanitation Ladder*, Vientiane, Laos.

24   Phil Davidson (1993) 'Cholera menace returns to Mexico', *The Independent*, 31 August, cited in John Pickford (1996) *Low-cost Sanitation: A Survey of Practical Experience*, IT Publications, London.

25   V. S. Naipaul (1981) *Among the Believers*, Knopf, New York.

26   George Gillanders (1940) 'Rural housing', *Journal of the Royal Institute of Sanitation*, vol 60, no 6, pp230–240.

27   WSP (2002) 'Selling sanitation in Vietnam: What works?', WSP field note, Water and Sanitation Program – East Asia and the Pacific, Jakarta, p10.

28   David Hall and Michael Adams (1991) 'Sanitation and hygiene practices', Chapter 4 in *Water, Sanitation, Hygiene and Health in the Qabane Valley, Lesotho*, for Tebellong

Hospital Primary Health Care Department, Lesotho.

29 Raymond Iseley, Scott Faiia, John Ashworth, Richard Donovan and James Thomson (1986) 'Framework and guidelines for CARE water supply and sanitation projects', WASH Technical Report No 40, WASH, Arlington, VA, cited in Pickford, op cit (Note 24).

30 Ramanantsoa, op cit (Note 22).

31 Maggie Black, personal experience.

32 Sarah Lloyd (1992) *An Indian Attachment*, Eland, London.

33 Maggie Black (1990) *From Handpumps to Health*, UNICEF, New York, p118.

34 P. Utiang Ugbe (1990) coursework at WEDC, Loughborough University, cited in Pickford, op cit (Note 24).

35 A. Agarwal (1981) *Water, Sanitation, Health – For All*, Earthscan, London, cited in Pickford, op cit (Note 24).

36 This account is based on research undertaken for one of the author's earlier books: Maggie Black with Rupert Talbot (2005) *Water, A Matter of Life and Health*, Oxford University Press India, New Delhi.

37 UNDP (2006) 'The vast deficit in sanitation', Chapter 3 in *Human Development Report*, UNDP, New York.

38 Maggie Black (2006) visit to the programme, October 2006.

39 The World Bank (2007) 'Bangladesh, country environmental analysis', The World Bank, Washington, DC, www.worldbank.org.bd/, accessed March 2007.

40 NGO Forum for Drinking Water Supply and Sanitation (2005) *WatSan Information Booklet*, NGO Forum for Drinking Water Supply and Sanitation, Dhaka, cited in Rokeya Ahmed (2005) 'Shifting millions from open defecation to hygienic practices', WaterAid and the Asian Development Bank, www.adb.org/Documents/Events/2005/Sanitation-Wastewater-Management/paper-ahmed.pdf, accessed 9 May 2007.

41 Black, op cit (Note 33), p18.

42 UNICEF and DPHE (1997) *Rivers of Change: New Directions in Sanitation, Hygiene and Water Supply in Bangladesh*, UNICEF and the Department of Public Health Engineering, Dhaka.

43 WSP (2007) 'Paradigm shifts: Total sanitation in South Asia', Water and Sanitation Program presentation, World Bank Sanitation Week, Washington, February 2007.

44 Sandy Cairncross (2006) oral evidence to the UK House of Commons International Development Committee, Ev 38, *Sanitation and Water, Vol II, Oral and Written Evidence*, HMG, London.

45 Kamal Kar (2003) 'Subsidy or self-respect? Participatory total community sanitation in Bangladesh', IDS Working Paper 184, Institute of Development Studies, University of Sussex, Brighton, UK.

46 UNICEF (2003) 'Sanitation, hygiene education and water supply in Bangladesh', UNICEF field note, Dhaka, citing National Baseline Survey statistics for 2003.

47   Kamal Kar and Petra Bongartz (2006) *Update on Some Recent Developments in CLTS*, Institute of Development Studies, University of Sussex, Brighton, UK.

48   Shafiul Azam Ahmed (undated) *Who Pays for Sanitation? Lessons from Bangladesh*, Water and Sanitation Program, Dhaka.

49   WSP, op cit (Note 43).

50   Kalinga Tudor Silva and Karunatissa Athukorala (1991) *The Watta-dwellers: A Sociological Study of Selected Urban Low-income Communities in Sri Lanka*, University Press of America, Lanham, MD, cited in Pickford, op cit (Note 24).

Chapter 4

1   Mariam Dossal (1991) *Imperial Designs and Indian Realities: The Planning of Bombay City 1845–1875*, Oxford University Press, Bombay.

2   Sjaak Van der Geest (2002) 'The night-soil collector: Bucket latrines in Ghana', *Postcolonial Studies*, vol. 5, no 2, pp197–206.

3   Isabel C. Blackett (1994) 'Low-cost urban sanitation in Lesotho', Water and Sanitation Discussion Paper Series, Water and Sanitation Program, The World Bank, Washington, DC, p3.

4   A. K. Roy, P. K. Chatterjee, K. N. Gupta, S. T. Khare, B. B. Rau and R. S. Singh (1984) 'Manual on the design, construction and maintenance of low-cost pour-flush waterseal latrines in India', TAG Technical Note Number 10, TAG-India, The World Bank, Washington, DC.

5   T. V. Luong, Ongart Chanacharnmongkol and Thira Thatsanatheb (2002) 'Universal sanitation in rural Thailand', *Waterfront*, no 15, UNICEF WES Division, New York, p8.

6   Ibid.

7   Steven Esrey, J. Gough, D. Rapaport, R. Sawyer, M. Simpson-Hébert, J. Vargas and U. Winblad (ed) (1998) *Ecological Sanitation*, Sida, Stockholm, p21.

8   WSP (2002) 'Selling sanitation in Vietnam: What works?', field note, Water and Sanitation Program East Asia and the Pacific, Jakarta.

9   Ibid, p28.

10   Ken Saro-Wiwa (1986) 'The inspector calls', in *A Forest of Flowers*, Saros International Publishers, Port Harcourt, Nigeria.

11   Van der Geest, op cit (note 2).

12   Andy Robinson (2002) 'VIP latrines in Zimbabwe: From local innovation to global sanitation solution', field note, Water and Sanitation Program, Africa Region, Nairobi.

13   Peter Morgan (2005) 'Zimbabwe's rural sanitation programme – An overview of the main events', unpublished document, Harare.

14   Maggie Black, personal communication with public health worker resident in Zimbabwe in the early 1980s.

15 Peter Morgan (1976) 'The pit latrine revived', *Central African Journal of Medicine*, no 23, pp1–4.

16 Duncan Mara (1996) *Low-Cost Urban Sanitation*, John Wiley and Sons, Chichester, UK.

17 Maggie Black (1998) *Learning What Works: A 20-Year Retrospective View on International Water and Sanitation Cooperation 1978–1998*, UNDP/World Bank Water and Sanitation Program, The World Bank, Washington, DC.

18 WSP (1990) 'Rural sanitation in Lesotho: From pilot project to national programme', UNDP/World Bank Water and Sanitation Program discussion paper, The World Bank, Washington, DC.

19 Ibid.

20 John van Nostrand and James G. Wilson (1983) 'Rural ventilated improved pit latrines: A field manual for Botswana', TAG Technical Note Number 8, The World Bank, Washington, DC.

21 WSP and PROWESS (1996) 'Toolkit: Good practice on gender in water and sanitation', The World Bank, Washington, DC, http://siteresources.worldbank.org/INTGENDER/Resources/toolkit.pdf.

22 Ian Pearson (2002) 'The National Sanitation Programme in Lesotho: How political leadership achieved long-term results', Water and Sanitation Program field note, The World Bank, Washington, DC.

23 Steve Anankum (1991) coursework at WEDC, Loughborough University, cited in John Pickford (1995) *Low-cost Sanitation: A Survey of Practical Experience*, IT Publications, London.

24 WSP, op cit (note 18), p15.

25 Barbara McCrea, Tony Pinchuk and Donald Reid (1997) *South Africa, Lesotho and Swaziland, The Rough Guide*, Rough Guides, London.

26 Morgan, op cit (note 13).

27 Comfort Olyiwole and Z. O. Agberemi (1998) 'Making SanPlats a household word', *Waterfront*, no 12, UNICEF WES Division, New York.

28 Dale Whittington, Donald Lauria and Albert Wright (1992) *Household Demand for Improved Sanitation Services: A Case Study of Kumasi, Ghana*, UNDP/World Bank Water and Sanitation Program, The World Bank, Washington, DC.

29 Morgan, op cit (note 13).

30 http://siteresources.worldbank.org/INTEEI/Data/20857101/Lesotho.pdf, accessed May 2007.

31 Maggie Black (2007) discussion with John Eichelstein, Idée Casamance, Senegal.

32 Pranab Shah (2007) 'Vault latrines: Traditional sanitation technology option used in Afghanistan', unpublished note from WASH specialist, UNICEF Afghanistan.

33 WSP (2002) 'The National Sanitation Programme in Mozambique: Pioneering peri-urban sanitation', field note, Water and Sanitation Program – Africa Region, The World Bank, Washington, DC.

34 Sandy Cairncross (1992) 'Sanitation and water supply: Practical lessons from the Decade', Water and Sanitation Discussion Paper No 9, Water and Sanitation Program, The World Bank, Washington, DC.

35 WSP, op cit (note 33).

36 www.sanplat.com/sanplat, accessed May 2007.

37 Maggie Black (2007) UNICEF-assisted field visit to Belgian aid-assisted rural sanitation programme, based at Bambey, March 2007.

38 www.sulabhinternational.org/pg15.htm, accessed May 2007.

39 Philip Wan (1988) *Sanitation Programme India, 1982–88: Reflections*, UNICEF, New Delhi.

40 Maggie Black with Rupert Talbot (2004) *Water: A Matter of Life and Health*, Oxford University Press India, New Delhi.

41 Ibid.

42 UNICEF (1995) *Watsan India 2000*, UNICEF, New Delhi.

43 Karan Manveer Singh (2007) 'Unrealistic approach killing rural sanitation programme', *Down to Earth*, Centre for Science and Environment, New Delhi, 15 April 2007, www.downtoearth.org.in/full6.asp?foldername=20070415& filename=news&sec_id=50&sid=25, accessed January 2008, accessed May 2007.

44 Interview with Maggie Black, November 2006.

45 P. Sainath (1996) *Everyone Loves a Good Drought: Stories from India's Poorest Districts*, Penguin India, New Delhi.

46 Andreas Schönborn (2000) 'Don't mix', *EcoEng Newsletter*, no 3, December, www.iees.ch/EcoEng003/EcoEng003_Int.html, accessed May 2007.

47 Witold Rybczynski, Chongrak Polprasert and Michael McGarry (1982) *Appropriate Technology for Water Supply and Sanitation: A State-of-the-art Review and Annotated Bibliography*, The World Bank, Washington, DC.

48 Steven Esrey, Ingvar Andersson, Astrid Hillers and Ron Sawyer (2001) *Closing the Loop: Ecological Sanitation for Food Security*, Publications on Water Resources No 18, UNDP, New York, and Sida, Stockholm.

49 Esrey et al, op cit (note 7).

50 Esrey et al, op cit (note 48).

51 These figures are from Sweden, cited in Maggie Black (2001) *Conference Report: First International Conference on Ecological Sanitation, Nanning, China*, Sida, Stockholm.

52 Louise Emilia Dellström Rosenquist (2005) 'A psychosocial analysis of the human-sanitation nexus', *Journal of Environmental Psychology*, no 25, pp335–346.

53 Data sheets for ecosan projects, GTZ Germany, www.gtz.de/de/dokumente/en-ecosan-pds-005-china-guanxi-2005.pdf, accessed May 2007.

54 Center for Innovation in Alternative Technologies website, www.laneta.apc.org/esac/citaing.htm, and Esrey et al, op cit (note 48).

55 Richard Holden and Aussie Austin (1999) 'Introduction of urine-diversion in South Africa', paper for the 25th WEDC International Conference, Addis Ababa,

Ethiopia, www.lboro.ac.uk/wedc/papers/25/039.pdf, accessed May 2007.

56 Programme of the MSWAS on www.gaisa-mspas.gob.sv, accessed May 2007.

57 WaterAid Uganda, www.wateraid.org/uganda/news/5461.asp.

58 Mayling Simpson-Hébert and Uno Winblad (eds) (2004) *Ecological Sanitation* (revised edition), Stockholm Environmental Institute, Stockholm.

59 Peter Morgan (2004) 'An ecological approach to sanitation: A compilation of experiences', unpublished document, Harare.

60 Jean-François Pinera and Lisa Rudge (2005) 'Water and sanitation assistance for Kabul: A lot for the happy few?', paper submitted to the 31st WEDC International Conference, Kampala, Uganda, http://wedc.lboro.ac.uk/conferences/pdfs/31/Pinera.pdf, accessed May 2007.

61 Edward D. Breslin (2003) 'Introducing ecological sanitation in northern Mozambique', WaterAid, www.wateraid.org/documents/introducing_eco_san_in_n_mozambique_2003.pdf, accessed May 2007.

Chapter 5

1 Sandy Cairncross (1992) 'Sanitation and water supply: Practical lessons from the Decade', Water and Sanitation Discussion Paper No 9, Water and Sanitation Program, The World Bank, Washington, DC.

2 Maggie Black (1990) *From Handpumps to Health*, UNICEF, New York.

3 Carol de Rooy and L. A. Donaldson (1990) 'Integrated water and sanitation development: The Nigerian Experience', *Water Quality Bulletin*, vol 15, no 1.

4 Maggie Black (1983) personal experience.

5 Maggie Black (1996) *Children First: The story of UNICEF*, UNICEF, New York, and Oxford University Press, Oxford, p115.

6 Y. D. Mathur (1998) 'The Clean Friday movement', *Waterfront*, no 12, UNICEF WES Division, New York.

7 Gareth Richards (2005) *Systematisation of the Friendly and Healthy School Initiative's School Sanitation and Hygiene Education Component*, UNICEF, Managua.

8 UNICEF (2006) *Schools as Partners in Sanitation and Hygiene*, Module B of International Learning Exchange programme materials, UNICEF, New Delhi.

9 IRC (2004) *The Way Forward: Construction is not Enough*, Proceedings of the School Sanitation and Hygiene Education Symposium, IRC, Delft, The Netherlands.

10 Guy Hutton and Laurence Haller (2004) *Evaluation of the Costs and Benefits of Water and Sanitation Improvements at the Global Level*, WHO Geneva.

11 IRC, op cit (note 9), presentation by Darren Saywell, Water Supply and Sanitation Collaborative Council, Geneva.

12 IRC, op cit (note 9).

13 See the research project of the London School of Hygiene and Tropical Medicine, www.hygienecentral.org.uk/proj_schoolsanitation.htm, accessed June 2007.

14 Sowmyaa Bharadwaj and Archana Patkar (2004) *Menstrual Hygiene and Management in Developing countries: Taking Stock*, Junction Social, Mumbai, India.

15 Ibid; case reported by Kalpravriksh, an NGO based in Pune.

16 2003 figures provided by UNICEF Jaipur, cited in Maggie Black with Rupert Talbot (2005) *Water: A Matter of Life and Health*, Oxford University Press India, New Delhi; accounts of SSHE in Tonk and Mysore also originally collected for this earlier book.

17 Sumita Ganguly (2007) 'Presentation on SSHE in India: An investment in children', UNICEF, New Delhi.

18 Maggie Black with Rupert Talbot (2005) *Water: A Matter of Life and Health*, Oxford University Press India, New Delhi, pp125–127.

19 P. Amudha (2007) *Scaling up School Sanitation and Hygiene Education with Quality*, Case Study IIR 2007, UNICEF, New Delhi.

20 Kochurani Mathew (2007) 'The School Health Clubs Project in Kerala', information note, UNICEF, New Delhi, www.schools.watsan.net/page/160/offset/10, accessed May 2007.

21 Rosemary Rop (2007) 'School health clubs: Can they change hygiene behaviour?', UNICEF information note, www.schools.watsan.net/page/160/offset/10, accessed May 2007.

22 Maggie Black (2006), information provided by UNICEF Nicaragua and collected in the field.

23 UNICEF (2004) *Building for Life: A Package to Achieve a Secure and Healthy Learning Environment*, UNICEF, Dakar, Senegal.

24 Figures provided by M. Lamine Bodian, chief of the Hydrological Division and interim chief of the Sanitation Bureau, Ziguinchor, Casamance, March 2007.

25 UNICEF (2007) 'Myanmar experiences and lessons learnt from the National Sanitation Week movement', information note provided by UNICEF Myanmar.

26 Mayling Simpson-Hébert, Ron Sawyer and Lucy Clarke (1996) *The PHAST Initiative: Participatory Hygiene and Sanitation Transformation: A New Approach to Working with Communities*, WHO, Geneva and UNDP/World Bank Water and Sanitation Program, Washington, DC.

27 Juliet Waterkeyn and Sandy Cairncross (2005) 'Creating demand for sanitation and hygiene through community health clubs: A cost-effective intervention in two districts in Zimbabwe', *Social Science and Medicine*, no 61, pp1958–1970.

28 Ibid.

29 Ibid.

30 Sandy Cairncross (2006) oral evidence to the UK House of Commons International Development Committee, Ev 38, *Sanitation and Water, Vol II, Oral and Written Evidence*, HMG, London.

31 WSP (2007) *From Burden to Communal Responsibility: A Sanitation Success Story from Southern Region in Ethiopia*, Water and Sanitation Program, The World Bank,

Washington, DC.

32 Ibid.

33 Marion W. Jenkins (2004) *Who Buys Latrines, Where and Why?*, Water and Sanitation Program field note, The World Bank, Washington, DC.

34 Marion W. Jenkins and Beth Scott (2007) 'Behavioural indicators of household decision-making and demand for sanitation and potential gains from social marketing in Ghana', *Social Science and Medicine*, vol 64, no 12, pp2427–2442.

35 www.sanplat.com/sanplat, accessed September 2007.

Chapter 6

1   Roberto Lenton, Albert M. Wright and Kristen Lewis (2005) *Health, Dignity and Development: What will it Take?*, Final Report of the UN Millennium Project Task Force, UNDP, New York, and Earthscan, London.

2   Mari Marcel Thekaekara (2005) 'Combating caste', *New Internationalist*, no 380.

3   This figure was quoted at a Supreme Court hearing in 2005 (see V. Venkatesan (2005) 'A case for human dignity', *Frontline*, The Hindu, Delhi, vol 22, no 12, 4–17 June); the official estimate of the Ministry of Social Justice and Empowerment for 2002/2003 was 676,000, up from 588,000 in 1992.

4   Deepa Joshi (2005) *Sanitation for the Urban Poor: Whose Choice, Theirs or Ours?*, Report of a DFID-funded research project, managed by the Institute of Irrigation and Development Studies, University of Southampton, Southampton, UK.

5   Annie Zaidi (2006) 'India's shame', cover story, *Frontline*, The Hindu, Delhi, vol 23, no 18, 9–22 September.

6   V. Venkatesan (2005) 'A case for human dignity', *Frontline*, The Hindu, Delhi, vol 22, no 12, 4–17 June.

7   Sowmyaa Bharadwaj and Archana Patkar (2004) *Menstrual Hygiene and Management in Developing Countries: Taking Stock*, Junction Social, Mumbai.

8   Mari Marcel Thekaekara (2002) *Endless Filth: The Saga of the Bhangis*, Books for Change, Bangalore, India.

9   Jo Beall (2002) 'Globalization and social exclusion in cities: Framing the debate with lessons from Africa and Asia', *Environment and Urbanization*, vol 14, no 1, pp41–51.

10 Articles by Annie Zaidi and V. Venkatesan in *Frontline*, op cit (notes 5 and 6).

11 S. P. Singh (2002) *Sulabh Sanitation Movement: Vision-2000 Plus*, Sulabh International Social Service Organisation, New Delhi

12 Gita Ramaswamy (2005) *India Stinking: Manual Scavengers in Andhra Pradesh*, Navayana Publishing, Chennai, India.

13 Sjaak Van Der Geest (1998) 'Akan shit: Getting rid of dirt in Ghana', *Anthropology Today*, vol 14, no 3, pp8–12.

14   Tova Maria Solo (1999) 'Small-scale entrepreneurs in the urban water and sanitation market', *Environment and Urbanization*, vol 11, no 1, pp117–132, IIED, London.

15   The World Bank (1992) *Development and the Environment*, The World Development Report 1992, The World Bank, Washington, DC.

16   Charlotte Denny (2003) 'The poor are not profitable and foreign firms are pulling the plug', *The Guardian*, 23 August.

17   Solo, op cit (note 14).

18   Tova Maria Solo (1998) 'Water and sanitation services for the poor: Profiles of small entrepreneurs', draft paper prepared for Water and Sanitation Program, The World Bank, Washington, DC.

19   WaterAid Tanzania (2003) 'Water reforms and PSP in Dar-es-Salaam', WaterAid and Tearfund, London, www.wateraid.org/documents/plugin_documents/ waterreformsandpsptanz.pdf, accessed May 2007.

20   WaterAid and Tearfund (2003) *New Rules, New Roles: Does PSP Benefit the Poor? Case Studies of Private Sector Participation in Water and Sanitation in 10 Countries*, WaterAid and Tearfund, London.

21   Bernard Collignon and Marc Vézina (2000) *Independent Water and Sanitation Providers in African Cities: Full Report of a Ten-country Study*, WSP with other partners, The World Bank, Washington, DC.

22   Racine Kane (2007) personal communication, UNICEF WASH consultant, UNICEF Dakar, March.

23   Christine Werner (2007) 'Ecosan experiences and the sustainable sanitation alliance', presentation from GTZ at International Sanitation Workshop hosted by Oxfam GB, Oxford, 20–21 September.

24   Maggie Black (1997) *Rivers of Change: New Directions in Sanitation, Hygiene and Water Supply in Bangladesh*, DPHE and UNICEF Bangladesh, Dhaka.

25   Maggie Black (1990) 'Bangladesh case study', in *From Handpumps to Health*, UNICEF, New York.

26   Unpublished note from UNICEF New Delhi on the market economy aspects of the Total Sanitation Campaign, June 2007.

27   Maggie Black with Rupert Talbot (2005) *Water: A Matter of Life and Health*, Oxford University Press India, New Delhi.

28   WSP (2005) *Understanding Small Scale Providers of Sanitation Services: A Case Study of Kibera*, Water and Sanitation Program, The World Bank, Washington, DC.

29   Maggie Black (2006) interview in the field, October.

30   See fuller discussion in Chapter 2, based on Martin Mulenga, Gift Manase and Ben Fawcett (2004) *Building Links for Improved Sanitation in Poor Urban Settlements*, Institute of Irrigation and Development Studies, University of Southampton, Southampton, UK.

31   Maria S. Muller (ed) (1997) *The Collection of Household Excreta: The Operation of Services in Urban Low-income Neighbourhoods*, WASTE, Gouda, The Netherlands, and ENSIC,

Asian Institute of Technology, Bangkok, Thailand.

32 BPD (undated) 'Sanitation partnerships: Dar-es-Salaam case study', BPD Sanitation Series, www.bpd-waterandsanitation.org/web/w/www_38_en.aspx, accessed February 2007.

33 Ibid.

34 Fred Pearce (2000) 'It's not just high-flown ideas...', *New Scientist*, vol 166, no 2240, 27 May, p14.

35 Discussion with Manus Coffey, June 2007.

36 Manus Coffey (2006) 'Off-site disposal of latrine wastes', proposed paper for Dakar Conference, May, received from the author.

37 Haroon Ur Rashid (2004) *Vacutug Service for Collection and Disposal in Dhaka; Decentralized Wastewater Management in Bangladesh, Case Study*, WaterAid Bangladesh, Dhaka.

38 Steven Sugden's 'gulper' is one example; presentation at an International Sanitation Workshop, hosted by Oxfam GB, Oxford, 20–21 September 2007.

39 Marion W. Jenkins and Steven Sugden (2006) 'Rethinking sanitation: Lessons and innovation for sustainability and success in the new millennium', Occasional Paper 2006/27, UN Human Development Report, New York.

40 Mike Davis (2006) *Planet of Slums*, Verso, London, pp71–72.

41 Jeremy Seabrook (1996) *In the Cities of the South: Scenes from a Developing World*, Verso, London, pp196–197.

42 Practical Action Consulting (2006) 'Bangladesh rural sanitation supply chain and employment impact', Occasional Paper 2006/43, UN Human Development Report, New York.

43 Sandy Cairncross (2004) 'The case for marketing sanitation', Water and Sanitation Program field note, The World Bank, Washington, DC.

44 Suzanne Hanchett, Shireen Akhter, Mohidul Hoque Khan, Stephen Mezulianik and Vicky Blagbrough (2003) 'Water, sanitation and hygiene in Bangladeshi slums: An evaluation of the WaterAid-Bangladesh urban programme', *Environment and Urbanization*, vol 15, no 2, pp43–56, IIED, London.

45 Gift Manase (2003) 'Cost recovery for sanitation services: The case of poor urban areas in Zimbabwe', PhD thesis submitted to Faculty of Engineering and Applied Sciences, University of Southampton, Southampton, UK.

46 Interviews with Dorcas Pratt, Gilbert Nkusi and staff of WaterAid, Antananarivo, and Father Edwin Joseph of FSG, March 2007.

Chapter 7

1 WHO and UNICEF (2006) *Meeting the MDG Drinking Water and Sanitation Targets: The Urban and Rural Challenge of the Decade*, WHO/UNICEF Joint Monitoring Programme, Geneva and New York; definitions are on p4.

2   Ibid, p21.

3   Ibid, p21.

4   Figures calculated from UNDESA (2007) *The Millennium Development Goals Report 2007*, United Nations Inter-Agency and Expert Group on MDG Indicators, New York.

5   Marion W. Jenkins and Steven Sugden (2006) 'Rethinking sanitation: Lessons and innovation for sustainability and success in the new millennium', Occasional Paper 2006/27, UN Human Development Report, New York.

6   House of Commons International Development Committee (2007) 'Sanitation and water', sixth report of Session 2006/07, vol 1, HMG, London.

7   Ibid, Appendix; DFID is issuing a new policy document on sanitation in late 2007.

8   IRC (2007) 'India: Unrealistic approach hampers rural sanitation programme', *IRC Source Weekly*, 21 July; quoting *Down to Earth* magazine, www.irc.nl/page/35966.

9   Conference Proceedings for the South Asian Conference on Sanitation (2004), Dhaka, cited in UN-Habitat (2006) *State of the World's Cities 2006/7*, UN-Habitat, Nairobi, and Earthscan, London, p83.

10 Nilanjana Mukherjee (2001) *Achieving Sustained Sanitation for the Poor: Policy and Strategy Lessons from Participatory Assessments in Cambodia, Indonesia and Vietnam*, Water and Sanitation Program – East Asia and the Pacific, Jakarta.

11 Sandy Cairncross, Steven Sugden, Mimi Jenkins and Beth Scott (LSTHM Sanitation Group) (2007) 'Sanitation: What we know and what we need to know', presentation at International Sanitation Workshop hosted by Oxfam GB, Oxford, 20–21 September.

12 WSP (2004) 'Who buys latrines, where and why?', Water and Sanitation Program field note, the World Bank, Washington, DC.

13 See discussion in Chapter 4; point made to Maggie Black in visit to villages in Yongning County in 2001.

14 Discussion between Maggie Black and Rolf Lyendijk of the WHO/UNICEF Joint Monitoring Programme Team, September 2007.

15 WHO and UNICEF, op cit (note 1).

16 Ibid, p4.

17 WSP (2004) *From Hazard to Convenience: Towards Better Management of Public Toilets in the City of Nairobi*, Water and Sanitation Program, The World Bank, Washington, DC.

18 John Pickford (1995) *Low-cost Sanitation: A Survey of Practical Experience*, IT Publications, London, p40.

19 Manus Coffey (2006) 'Off-site disposal of latrine wastes', proposed paper for Dakar Conference, May, unpublished, received from the author.

20 Cranfield University, Aguaconsult and IRC (2006) 'Landscaping of technologies, Annex 13: Bio-additives', paper for the Bill and Melinda Gates Foundation, Cranfield, UK.

21 See, for example, www.ecosanres.org, www.sanplat.com, www.sandec.ch, http://aquamor.tripod.com, www.lshtm.ac.uk/dcvbu.

22 Steven Sugden (2006) 'The microbiological contamination of water supplies', WELL factsheet, www.lboro.ac.uk/well/resources/fact-sheets/fact-sheets-htm/Contamination.htm, accessed June 2007.

23 Maggie Black with Rupert Talbot (2005) *Water: A Matter of Life and Health*, Oxford University Press India, New Delhi, p114.

24 Jenkins and Sugden, op cit (note 5).

25 Rakotoniana Patrice (2000) 'Document de synthèse sur le sous-secteur de l'assainissement', Government of Madagascar, quoted in Government of Madagascar, WaterAid and UNICEF (2003) *Assainissement, le Défi*, Antananarivo.

26 David Redhouse (2005) *Getting to Boiling Point: Turning up the Heat on Water and Sanitation*, WaterAid, London.

27 Jenkins and Sugden, op cit (note 5).

28 Gert de Bruijne (2007) Presentation by WASTE at International Sanitation Workshop hosted by Oxfam GB, Oxford, 20–21 September.

29 WaterAid (2006) 'Written evidence to the House of Commons International Development Committee', www.publications.parliament.uk/pa/cm200607/cmselect/cmintdev/ucwater/uc3002.htm, accessed June 2007.

30 GWP (2000) *Towards Water Security: A Framework for Action*, The Global Water Partnership, Stockholm.

31 WaterAid, op cit (note 29).

32 Guy Hutton and Laurence Haller (2004) *Evaluation of the Costs and Benefits of Water and Sanitation Improvements at the Global Level*, WHO, Geneva.

33 World Bank (2006) 'Annual Report 2006', The World Bank, Washington, DC, http://web.worldbank.org/WBSITE/EXTERNAL/EXTABOUTUS/EXTANNREP, accessed June 2007.

34 Sunita Narain, Suresh Babu SV and Bharat Lal Seth (2007) *Sewage Canal: How to Clean the Yamuna*, Centre for Science and Environment, New Delhi.

35 WSSCC (2006) 'For her, it's the big issue, putting women at the centre of water supply, sanitation and hygiene', evidence report by the Water Supply and Sanitation Collaborative Council, Geneva.

36 WEDC (2004) 'Water and environmental sanitation – Working towards equity and inclusion for disabled people: A discussion paper', WEDC, Loughborough University, Loughborough, UK.

37 UNDP (2006) *Human Development Report: Beyond Scarcity: Power, Poverty and the Global Water Crisis*, UNDP, New York, p114.

38 Profile of Doly Akter compiled in 2006 by UNICEF Bangladesh.

Index

Printed and bound by CPI Group (UK) Ltd, Croydon, CR0 4YY

27/10/2024

01779947-0003